AF361968

Composites from Recycled and Modified Woods – Technology, Properties, Application

Composites from Recycled and Modified Woods – Technology, Properties, Application

Editors

Ladislav Reinprecht
Ján Iždinský

MDPI • Basel • Beijing • Wuhan • Barcelona • Belgrade • Manchester • Tokyo • Cluj • Tianjin

Editors
Ladislav Reinprecht
Department of Wood Technology
Technical University in Zvolen
Zvolen
Slovakia

Ján Iždinský
Department of Wood Technology
Technical University in Zvolen
Zvolen
Slovakia

Editorial Office
MDPI
St. Alban-Anlage 66
4052 Basel, Switzerland

This is a reprint of articles from the Special Issue published online in the open access journal *Forests* (ISSN 1999-4907) (available at: https://www.mdpi.com/journal/forests/special_issues/ antimicrobial_wood).

For citation purposes, cite each article independently as indicated on the article page online and as indicated below:

LastName, A.A.; LastName, B.B.; LastName, C.C. Article Title. *Journal Name* **Year**, *Volume Number*, Page Range.

ISBN 978-3-0365-3109-0 (Hbk)
ISBN 978-3-0365-3108-3 (PDF)

Contents

Preface to "Composites from Recycled and Modified Woods – Technology, Properties, Application"

Wood composites are often used in furniture products and building structures. This book deals with preparation of particleboards (PBs), laminated veneer lumbers (LVLs), plywood, and medium-density fiberboards (MDFs) from recycled and modified woods, using traditional or new types of adhesives and additives, together with the basic properties and applications of such prepared composites. First of all, we would like to thank all the involved authors for the valuable articles that enriched this book. In the first part of this book, the contributions focus on the physical, mechanical and biological properties of PBs made from: old PBs present in recycled furniture, new damaged PBs, old spruce pallets, and thermally modified woods. The second part is devoted to using heterogeneous wood veneers of different quality in the manufacturing of maximized-performance LVL panels by veneer grading and optimized positioning. The third section examines the penetration process of different polarity and molecular weight liquids through the surface of MDF and PB panels. The fourth part deals with the effect of birch and black alder veneer densification on the mechanical properties of plywood composed of layers of densified and non-densified veneers. The fifth section is devoted to the ecofriendly PBs and MDFs produced from industrial waste fibers and magnesium lignosulfonate adhesive. The sixth section is devoted to the carbonization effects of a CO_2 laser when modifying wood surfaces, in the native state or after treatment with polyvinyl acetate or polyurethane polymers, with worsened adhesion and increased mold resistance. In the seventh part, new environmentally friendly fire retardants for cellulose and cellulose-modified materials are discussed.

Ladislav Reinprecht, Ján Iždinský
Editors

Editorial

Composites from Recycled and Modified Woods—Technology, Properties, Application

Ladislav Reinprecht and Ján Iždinský

Department of Wood Technology, Faculty of Wood Sciences and Technology, Technical University in Zvolen, T. G. Masaryka 24, 960 01 Zvolen, Slovakia; reinprecht@tuzvo.sk
* Correspondence: jan.izdinsky@tuzvo.sk

Citation: Reinprecht, L.; Iždinský, J. Composites from Recycled and Modified Woods—Technology, Properties, Application. *Forests* **2022**, *13*, 6. https://doi.org/10.3390/f13010006

Received: 14 December 2021
Accepted: 17 December 2021
Published: 21 December 2021

Publisher's Note: MDPI stays neutral with regard to jurisdictional claims in published maps and institutional affiliations.

The intention of efficient processing and use of less valuable wood species, bio-damaged logs, sawmill residues, cuttings, chips, sawdust, recycled wooden products, and other lignocellulosic raw materials in the production of wood composites is the focus of several scientific research institutes in the world. Wood composites are mostly produced for a particular application and, therefore, the raw materials, needed additives, production processes, as well as finishing and surface treatments are adapted for these purposes. Research into the optimization of material and technological parameters of the production of wood composites with special properties and applications is still ongoing.

Recycled woods obtained from old buildings, furniture, industrial products, etc., as well as from modified woods prepared by thermal, chemical, or biological processes, have the potential to be a base or complementary raw material for construction and decorative wood composites used mostly in buildings, furniture, and for transport. Typical construction composites are glued prisms and boards (e.g., glulam, blockboard), glued large-area boards from veneers (e.g., plywood, laminated veneer lumber (LVL)), or large-area boards from wood particles and fibers (e.g., particleboard (PB), oriented strand board (OSB), medium density fiberboard (MDF)). More of the large-area boards can obtain a better decorative function usable in furniture and building architecture after veneering, lamination, coating, plasma-treating or other surface-treating technologies.

Within this Special Issue, selected articles related primarily to wood composites prepared from recycled and modified woods are collected in focus on the type and properties of used raw materials, additives, technological processes, including finishing and surface treatments, with an impact on the resulting properties and service-life of wood composites under their different uses. Development of wood and cellulosic composites with specific compositions and properties, coming out from the knowledge of wood material science, timber engineering, physic, and other areas of research, is also to interest of this Special Issue.

The Special Issue comprises ten articles by authors from nine countries—eight from Europe (Bulgaria, France, Germany, Greece, Poland, Romania, Slovakia, and Ukraine) and one from Asia (Iran). These articles represent a wide range of aspects related to: (1) material composition, e.g., species, amount, fraction, and distribution of recycled and modified woods, type of glues, biocides, fire retardants, and other additives; (2) processing and finishing; (3) properties of wood composites for construction and decorative use.

Authors in articles reported about: (a) new sources of the raw material for PBs, such as recycled wood waste generated in wood processing from faulty PBs bonded with urea-formaldehyde (UF) resin or taken from old furniture and other end-of-life wood products, including spruce pallets and thermally modified wood, taking into account the environmental and economic aspects of wood recyclates and obtaining the knowledge that the moisture, strength, and biological properties of PBs can be specifically improved, e.g., the resistance of PBs to swelling in water and to fungal decay increased at using particles from faulty PBs or thermowood cuttings [1–3]; (b) enhancing the use of heterogeneous wood veneers in the manufacturing of maximized-performance LVL panels by veneer grading and optimized positioning, as well as the LVL material mechanical property modeling, with a proposed optimization strategy of LVL manufacturing from variable-quality veneers in terms of their optimal stiffness or strength [4,5]; (c) the penetration process of liquids with different polarities and molecular weights through the surface of MDF and PB panels, when the most polar water molecules created strong and stable bonds with the cellulose of wood composites, and therefore, the penetration of water in their structure was the slowest [6]; (d) the effect of birch and black alder veneers' densification temperature and increased density on the mechanical plywood properties made from alternate layers of densified and non-densified veneers, e.g., the increase in the veneer's densification temperature from 150 to 210 °C resulted in a gradual decrease in the plywood's bending strength [7]; (e) the ecofriendly PBs and MDF composites produced from industrial waste fibers and magnesium lignosulfonate adhesive, at which such prepared composites had good mechanical properties but with worsened moisture properties [8]; (f) a CO_2 laser modification of beech and spruce wood surfaces before or after their coating with polyvinyl acetate (PVAc) or polyurethane (PUR) polymers, where the laser beams degraded and carbonized the wood adherent or the synthetic polymer layer and thus worsened the adhesion strength between wood and coating, while the mold resistance of laser-modified surfaces specifically increased [9]; (g) new fire retardants for cellulose and cellulose-modified materials, e.g., the expandable graphite, could be a very effective intumescent and environmentally friendly fire retardant [10].

Acknowledgments: We would like to thank MDPI *Forests*, as well as the Slovak Research and Development Agency under contract no. APVV-17-0583, for support of this editorial article.

Conflicts of Interest: The authors declare no conflict of interest.

References

1. Iždinský, J.; Vidholdová, Z.; Reinprecht, L. Particleboards from Recycled Wood. *Forests* **2020**, *11*, 1166. [CrossRef]
2. Iždinský, J.; Vidholdová, Z.; Reinprecht, L. Particleboards from Recycled Thermally Modified Wood. *Forests* **2021**, *12*, 1462. [CrossRef]
3. Iždinský, J.; Reinprecht, L.; Vidholdová, Z. Particleboards from Recycled Pallets. *Forests* **2021**, *12*, 1597. [CrossRef]
4. Duriot, R.; Pot, G.; Girardon, S.; Roux, B.; Marcon, B.; Viguier, J.; Denaud, L. New Perspectives for LVL Manufacturing from Wood of Heterogeneous Quality—Part. 1: Veneer Mechanical Grading Based on Online Local Wood Fiber Orientation Measurement. *Forests* **2021**, *12*, 1264. [CrossRef]
5. Duriot, R.; Pot, G.; Girardon, S.; Denaud, L. New Perspectives for LVL Manufacturing from Wood of Heterogeneous Quality—Part 2: Modeling and Manufacturing of Variable Stiffness Beams. *Forests* **2021**, *12*, 1275. [CrossRef]
6. Taghiyari, H.R.; Majidi, R.; Arsalan, M.G.; Moradiyan, A.; Militz, H.; Ntalos, G.; Papadopoulos, A.N. Penetration of Different Liquids in Wood-Based Composites: The Effect of Adsorption Energy. *Forests* **2021**, *12*, 63. [CrossRef]
7. Salca, E.-A.; Bekhta, P.; Seblii, Y. The Effect of Veneer Densification Temperature and Wood Species on the Plywood Properties Made from Alternate Layers of Densified and Non-Densified Veneers. *Forests* **2020**, *11*, 700. [CrossRef]
8. Antov, P.; Mantanis, G.I.; Savov, V. Development of Wood Composites from Recycled Fibres Bonded with Magnesium Lignosulfonate. *Forests* **2020**, *11*, 613. [CrossRef]
9. Reinprecht, L.; Vidholdová, Z. The Impact of a CO_2 Laser on the Adhesion and Mold Resistance of a Synthetic Polymer Layer on a Wood Surface. *Forests* **2021**, *12*, 242. [CrossRef]
10. Mazela, B.; Batista, A.; Grześkowiak, W. Expandable Graphite as a Fire Retardant for Cellulosic Materials—A Review. *Forests* **2020**, *11*, 755. [CrossRef]

 *forests*

Article

Particleboards from Recycled Wood

Ján Iždinský, Zuzana Vidholdová and Ladislav Reinprecht

Department of Wood Technology, Faculty of Wood Sciences and Technology, Technical University in Zvolen, T. G. Masaryka 24, 960 01 Zvolen, Slovakia; zuzana.vidholdova@tuzvo.sk (Z.V.); reinprecht@tuzvo.sk (L.R.)
* Correspondence: jan.izdinsky@tuzvo.sk

Received: 14 October 2020; Accepted: 30 October 2020; Published: 31 October 2020

Abstract: The effective recovery of wood waste generated in wood processing and also at the end of wood product life is important from environmental and economic points of view. In a laboratory, 16 mm-thick three-layer urea–formaldehyde (UF)-bonded particleboards (PBs) were produced at 5.8 MPa and 240 °C and with an 8 s/mm pressing factor, using wood particles prepared from (1) fresh spruce wood (C), (2) a mixture of several recycled wood products (R1), and (3) recycled faulty PBs bonded with UF resin (R2). Particles from spruce wood were combined with particles from R1 or R2 recyclates in weight ratios of 100:0, 80:20, 50:50 and 0:100. In comparison to the control spruce PB, the PBs containing the R1 recyclate from old wood products were characterized by lower thickness swelling after 2 and 24 h (TS-2h and TS-24h), lower by 18 and 31%; water absorption after 2 and 24 h (WA-2h and WA-24h), lower by 33 and 28%; modulus of rupture in bending (MOR), lower by 28%; modulus of elasticity in bending (MOE), lower by 18%; internal bond (IB), lower by 33%; and resistance to decay determined by the mass loss under the action of the brown-rot fungus *Coniophora puteana* (Δm), lower by 32%. The PBs containing the R2 recyclate from faulty PBs were also characterized by a lower TS-2h and TS-24h, lower by 45% and 59%; WA-2h and WA-24h, lower by 61% and 51%; MOR, lower by 37%; MOE, lower by 17%; and IB, lower by 33%; however, their biological resistance to *C. puteana* was more effective, with a decreased Δm in the decay test, lower by 44%.

Keywords: particleboards; recycled wood; physical and mechanical properties; decay resistance

---

1. Introduction

Today, manufacturing agglomerated materials, including particleboards (PBs), is based on the idea of utilizing the wood waste of lower value resulting from wood processing, such as sawdust, chips, particles, and wood pulp [1]. An increase in the volume of PB manufacturing has resulted in manufacturers seeking new material resources. At the present time, besides natural wood, other lignocellulose materials [2–4], as well as recycled wood [5,6], are used, especially in countries where there is a scarcity of wood resources [7].

Recycled wood is defined as various kinds of old and residual wood, such as wastes from furniture, construction, etc., and packaging as stated in the Waste Framework Directive (2008/98/EC) [6,8,9]. The increase in the volume of PB manufacturing is associated with PB consumption growth. In 2017, the manufacturing of PBs in Europe was 43.9×10^6 m^3, and in 2018, there was an increase of more than 4.1% [10,11]. Therefore, taking into account the continued growth in the production as well as in the consumption of PBs, it can be assumed that a large amount of materials from PBs will have to be eliminated or recycled each year [12]. Wan et al. [13] mentioned that there have been developed several methods for reconstituting the waste from PBs and other medium density fiberboards (MDF) in order to solve the problem of recycling the wood composites. The processing can be chemically thermomechanical [14,15], hydrothermal [16–18], chemical [19] or mechanical [20–24]. According to Wan et al. [13], it was confirmed by research results that common types of urea–formaldehyde (UF) and phenol–formaldehyde (PF) resins are suitable for manufacturing PBs and MDF from recycled wood materials using conventional technology.

According to Czarnecki et al. [24], the mentioned methods of wood waste reconstitution are rather difficult to apply in practice, besides the way of mechanical processing when manufacturing PBs. For example, the hydrolytic way of obtaining wood particles from residual PBs can only be used in the case of them being bonded with UF resin.

The wood waste from recycled composite materials should not be found in municipal waste landfills because of its possible degradation products responsible for damaging the environment [25–28]. It should be used as a material for new composites to become more environmentally friendly [25,27,29–34].

When using the recycled wood for composite manufacturing, the decontamination and subsequent grading can be considered a disadvantage. At the present time, several companies possess sophisticated equipment and technology for collecting, grading and decontaminating old timber. Subsequently, it can be used to manufacture PBs [35]. The properties of the new PB can be significantly affected by the recycled wood, owing to the type, size and additives used.

Several research studies aimed at determining the effects of various types of recycled wood and other lignocellulosic waste on the properties of PBs have been conducted across the world. Laskowska and Mamiński [36] studied the properties of PBs manufactured from plywood waste. When manufacturing three-layer PBs, Czarnecki et al. [24] replaced, in the core layer, 10 to 60% of the pine wood particles with particles produced from: (a) three-layer PBs bonded with UF resin, (b) PBs bonded with PF resin, and (c) MDF bonded with UF resin. They found out that the bending strength declined slightly when 10 to 50% particles obtained from recycled PBs bonded with UF resin were added. However, there was a significant decline in the case of 60% recycled particles being added. On the contrary, the internal bond (IB) declined significantly when the portion of recycled particles in the PBs was only 10%. The thickness swelling (TS) of PBs with higher portions of recyclates decreased after 2 as well as 24 h. PBs manufactured from particles obtained from recycled PBs bonded with PF resin showed the worst properties. This could be caused by the negative interaction of the original PF resin and new UF resin used. In this case, only PBs with a minimum of 10% recyclates in the central layer met the requirements of the standard. The best results were obtained using the particles from recycled MDF bonded with UF resin, whose 60% addition to PBs resulted in only a slight decrease in the modulus of rupture (MOR), and even an increase in the IB and TS after 2 and 24 h. Azambuja et al. [37] manufactured PBs using four types of wood recyclates from demolished buildings, i.e., MDF boards, PBs, plywood and timber. In comparison to control PBs from pine particles, the best properties were shown for the PBs manufactured from recycled timber. However, considering the low values of the MOR and modulus of elasticity (MOE), using the wood recyclates only in the core layer of PBs was recommended. Weber and Iwakiri [38] found out that plywood, MDF and PB wastes can be used to manufacture PBs, using them individually or in mixtures. Hameed et al. [39] manufactured three-layered PBs using two types of recycled wood waste material (untreated wood—waste made of massive wood—and slightly treated wood—waste issued from coating or gluing treatments and made of massive wood or other wood-based panels), which were bonded by a rape resin based on leftover cakes of rape oil in a natural state. Both PB types with recycled wood met the requirements of particleboard type P2 (no load-bearing panel for interior use in dry conditions) according to EN 312 [40].

The effects of two types of recycled wood particles – recyclates from old used wood products (R1) and from new but faulty PBs (R2), added in various amounts to spruce particles prepared from freshly cut logs—on selected physical, mechanical and biological properties of PBs were studied for the paper.

2. Materials and Methods

2.1. Materials

2.1.1. Wood Particles

By the company Kronospan Zvolen, Slovakia, industrial wood chips were prepared from (a) fresh spruce logs (C); (b) a mixture of several recycled wood products consisting of approximately 35% hardboards (HB) and MDF boards, 30% PBs, 20% pallets from spruce wood and 15% old furniture (R1); and (c) recycled faulty PBs bonded with UF resin (R2). All chip types were prepared in the Knife Ring Flakers G24 (GOOS Engineering s.r.o., Slovakia). Subsequently, from the chips were prepared wood particles in the laboratories of the Technical University in Zvolen, using the grinding mill SU 1 (TMS Pardubice, Czech Republic). The dimensions of the particles used in the core layer of the PBs ranged from 0.25 to 4.0 mm and in the surface layers from 0.125 to 1.0 mm. The oversized fraction and the dust were sorted and excluded from the experiment. Fractions falling through a sieve with a mesh diameter of 0.125 mm were considered to be the dust. Subsequently, the particles were dried to a moisture content of 2% in the case of the core layer and to a moisture content of 4% in the case of the surface layers.

2.1.2. Resin and Additives

The UF resin KRONORES CB 4005 D was used in the surface layers, and the UF resin KRONORES CB 1637 D, in the core layer of the PBs (Table 1). The UF resin was added to particles for the surface layers in an amount of 11%, and to particles for the core layer in an amount of 7%. Ammonium nitrate as a 57% water solution (2 or 4% of the dry mass of the UF resin in the case of the surface particles or core particles) was used as a hardener for the UF resins. A paraffin emulsion, with 35% dry mass, was applied on the surface and core particles in amounts of 0.6 and 0.7%, respectively.

Table 1. Properties of urea–formaldehyde (UF) resins.

Quality Parameters	Unit	Method	KRONORES CB 4005 D	KRONORES CB 1637 D
Solid content	%	EN 827 [41]	65.87	67.35
Ford cup viscosity, 4 mm/20 °C	s	EN ISO 2431 [42]	76	86
pH value		EN 1245 [43]	9.08	8.62
Gel time at 100 °C	s	Kronospan chloride test	81	36

2.2. Particleboard Preparation

Three-layer PBs with dimensions of $400 \times 300 \times 16$ mm and with a density of 650 ± 10 kg·m^{-3} were produced under laboratory conditions. UF resin with the hardener was applied on the conditioned wood particles in a laboratory rotary mixing device (VDL, TU Zvolen, Slovakia). The moisture content of the wood particles after applying the adhesive was 9.1 to 10.4% for the surface layers and 6.3 to 7.1% for the core layer. The particle mat was layered manually in wooden forms. The surface/core particle ratio was 35:65. The particle mat was cold pre-pressed in a low-temperature environment at a pressure of 1 MPa, and then, it was pressed in the CBJ 100-11 laboratory press (TOS, Rakovník, Czech Republic). Pressing was conducted in accordance with the pressing diagram (Figure 1)—at a maximum temperature of 240 °C, a maximum pressing pressure of 5.75 MPa, and a pressing factor of 8 s/mm. In total, 42 PBs were manufactured, i.e., 6 from each type (Table 2).

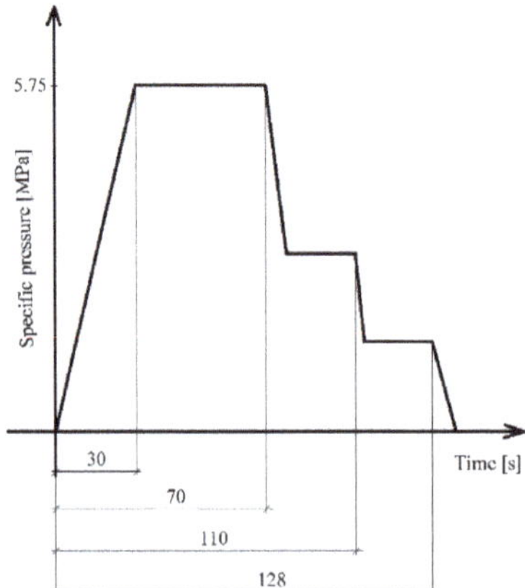

Figure 1. Standard three-stage pressing diagram for the manufacturing of particleboards (PBs).

Table 2. Individual types of manufactured particleboards (PBs).

Variant	Amount of Recycled Wood in PB, *w/w* (%)	Number of Produced Boards	Board Type
PB-C:			
100% particles of spruce wood	0	6	C
PB-R1:			
20%, 50% or 100% particles from	20	6	20 R1
mixture of recycled wood products,	50	6	50 R1
combined with 80, 50 or 0%	100	6	100 R1
particles of spruce wood			
PB-R2:			
20%, 50% or 100% particles from	20	6	20 R2
faulty PBs, combined with 80, 50	50	6	50 R2
or 0% particles of spruce wood	100	6	100 R2

2.3. Physical and Mechanical Properties of PBs

The selected properties of the PBs were determined according to the European (EN) standards or Slovak (STN) standards: the density by EN 323 [44], the moisture content by EN 322 [45], the TS and water absorption (WA) after 2 and 24 h by EN 317 [46] and STN 490164 [47], the MOR in bending and the MOE in bending by EN 310 [48], and the IB, i.e., the tensile strength perpendicular to the plane of the PBs, by EN 319 [49]. The samples for these tests were prepared from the 4-week air-conditioned PBs (Figure 2). The universal machine TiraTest 2200 (VEB TIW Rauenstein, Germany) was used to analyze the mechanical properties of the PBs. The classification of the PBs was performed in accordance with the European standard EN 312 [40], taking into account the requirements of the PB (type P2), with a thickness ranging between 13 and 20 mm.

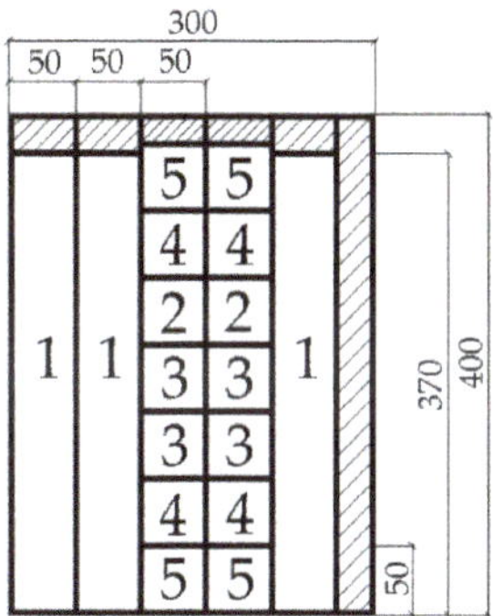

Figure 2. Cutting scheme for laboratory-manufactured PBs: 1—samples for testing modulus of rupture (MOR) and modulus of elasticity (MOE) by 3-point bending test [48], 2—samples for testing thickness swelling (TS) and water absorption (WA) after 2 and 24 h [46,47], 3—samples for testing internal bond (IB) [49] and density [44], 4—samples for testing biological resistance to the fungus *Coniophora puteana* [50], and 5—spare samples.

2.4. Decay Resistance of PBs

The resistance of all the prepared PBs to the brown-rot fungus *Coniophora puteana* (strain BAM Ebw. 15) was tested according to ENV 12038 [50] using the samples with dimensions of $50 \times 50 \times 16$ mm. Firstly, the samples were conditioned at a temperature of 20 ± 2 °C and a relative humidity of $65 \pm 2\%$, achieving an equilibrium moisture content (EMC) of $12 \pm 1\%$. The edges of the samples (50×16 mm) were sealed with the epoxy resin CHS-Epoxy 1200 mixed with the hardener P11 in a weight ratio of 11:1 (Stachema, Mělník, Czech Republic), in the amount of 200 ± 10 g·m^{-2}. Subsequently, the samples were oven dried at rising temperatures from 60 to 103 ± 2 °C for 10 h. The sterilized samples were cooled in a desiccator and weighed (m_0). Finally, all the surfaces of the PB samples were sterilized twice for 30 min with a UV light radiator. Then, they were soaked in distilled water for 240 min to achieve a moisture content of 25 to 30%. Finally, they were placed into 1 L Kolle flasks on top of the stainless steel grids, with the fungal mycelium inoculant grown on the agar-malt soil (HiMedia Laboratories Pvt. Ltd., Mumbai, India). All the Kolle flasks were incubated for 16 weeks at a temperature of 22 ± 2 °C and a relative humidity of 75 to 80%. After the mycological test, the samples of PBs—after depriving the fungal mycelia of their surfaces—underwent a gradual drying process until constant weights ($m_{0/decayed/}$) were achieved, i.e., 100 h at 20 °C, 1 h at 60 °C, 1 h at 80 °C, 8 h at 103 ± 2 °C, and final cooling in desiccators. Their mass losses were calculated in percentages using Equation (1):

$$\Delta m = \frac{m_0 - m_{0/decayed/}}{m_0} \cdot 100 \; (\%) \tag{1}$$

2.5. Statistical Analyses

The statistical software STATISTICA 12 was used to analyze the gathered data. The descriptive statistics deal with the basic statistical characteristics of the studied properties—the arithmetic mean and standard deviation. Simple linear correlation analysis together with the coefficient of determination was used as a method of inductive statistics to evaluate the measured data.

3. Results and Discussion

3.1. Physical and Mechanical Properties of PBs

The basic physical and mechanical properties of the PBs manufactured in the laboratory of TU in Zvolen are presented in Tables 3 and 4. The effects of the recyclates R1 and R2 on the properties of the PBs were analyzed by the linear correlations and the coefficients to determine r^2, as shown in Figures 3–7.

The density of the PBs ranged between 649 and 658 kg·m^{-3}. It was in a very narrow interval, and it was not apparently affected by the type and amount of wood recyclates (Tables 3 and 4). This finding was also confirmed by the r^2 values being lower than 0.01 (Figure 3). This means that the tested moisture, mechanical and biological properties of the PBs could not be affected by density.

Table 3. Physical and mechanical properties of the control PB (PB-C) and of the PBs containing particles from a mixture of wood recycled products—R1.

Property of PB		Recyclate R1 in PB *w/w* (%)			
		0	20	50	100
Density	(kg·m^{-3})	653 (18.22)	652 (18.42)	649 (23.81)	654 (22.26)
Thickness swelling (TS) after 2 h	(%)	6.05 (0.53)	4.98 (0.51)	4.93 (0.56)	5.65 (1.18)
Thickness swelling (TS) after 24 h	(%)	23.95 (1.39)	16.57 (2.64)	22.76 (3.52)	21.12 (3.83)
Water absorption (WA) after 2 h	(%)	27.59 (2.05)	18.54 (1.74)	21.38 (2.42)	19.55 (3.87)
Water absorption (WA) after 24 h	(%)	68.52 (2.33)	49.23 (3.95)	62.02 (6.05)	54.86 (6.52)
Internal bond (IB)	(MPa)	0.82 (0.06)	0.75 (0.04)	0.58 (0.06)	0.55 (0.06)
Modulus of rupture (MOR)	(MPa)	14.7 (1.57)	14.7 (1.48)	11.6 (1.33)	10.6 (1.40)
Modulus of elasticity (MOE)	(MPa)	2637 (288)	2666 (246)	2442 (170)	2155 (284)

Notes: Mean values: density from 42 samples, TS from 12 samples, WA from 12 samples, IB from 24 samples, MOR and MOE from 18 samples. Standard deviations are in the parentheses.

Table 4. Physical and mechanical properties of the control PB (PB-C) and of the PBs containing particles from recycled PBs—R2.

Property of PB		Recyclate R2 in PB *w/w* (%)			
		0	20	50	100
Density	(kg·m^{-3})	653 (18.22)	658 (16.53)	652 (20.10)	652 (21.63)
Thickness swelling (TS) after 2 h	(%)	6.05 (0.53)	4.48 (0.40)	4.19 (0.56)	3.32 (0.48)
Thickness swelling (TS) after 24 h	(%)	23.95 (1.39)	13.29 (1.35)	12.43 (1.26)	9.72 (0.98)
Water absorption (WA) after 2 h	(%)	27.59 (2.05)	16.41 (0.95)	15.35 (0.95)	10.80 (0.63)
Water absorption (WA) after 24 h	(%)	68.52 (2.33)	43.43 (1.91)	41.30 (1.51)	33.46 (1.44)
Internal bond (IB) strength	(MPa)	0.82 (0.06)	0.74 (0.05)	0.68 (0.05)	0.55 (0.12)
Modulus of rupture (MOR)	(MPa)	14.7 (1.57)	14.4 (1.18)	11.8 (1.24)	9.3 (1.27)
Modulus of elasticity (MOE)	(MPa)	2637 (288)	2800 (166)	2486 (173)	2194 (264)

Notes: Mean values: density from 42 samples, TS from 12 samples, WA from 12 samples, IB from 24 samples, MOR and MOE from 18 samples. Standard deviations are in the parentheses.

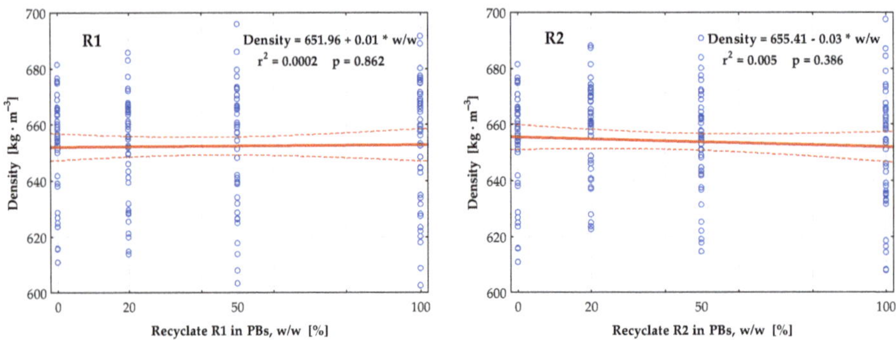

Figure 3. Density of PBs containing different types (R1 or R2) and amounts of wood recyclates.

The moisture properties of the PBs, i.e., the thickness swelling (TS) and water absorption (WA), were already more scattered: TS-2h ranged from 3.32% for PB-R2-100 to 6.05% for PB-C; TS-24h ranged from 9.72% for PB-R2-100 to 23.95% for PB-C; WA-2h ranged from 10.80% for PB-R2-100 to 27.59% for PB-C; WA-24h ranged from 33.46% for PB-R2-100 to 68.52% for PB-C (Tables 3 and 4). The effect of the

R1 recycled particles—from the mixture of recycled wood products—on the moisture properties of the PBs cannot be considered as significant, as the r^2 value of the linear correlations ranged only between 0.0001 and 0.09. Only in the case of WA-2h was the effect of the R1 particles slightly stronger, as the r^2 was 0.22 (Figures 4 and 5). The result that the larger decrease in TS and WA was observed in the case of the PBs that contained only minimal amounts (20%) of the R1 recyclate is interesting—TS-2h decreased by about 18%, TS-24h by about 31%, WA-2h by about 33%, and WA-24h by about 28%. On the contrary, the R1 recycled particles—from the faulty recycled PBs—significantly improved the moisture properties of the laboratory-prepared PBs, the most for PB-R2-100, containing 100% R2 recyclate, with decreases in TS-2h and TS-24h of about 45 and 59%, or in WA-2h and WA-24h of about 61 and 51% (Tables 3 and 4). This result can be explained by the presence of the cured UF resin macromolecules on the surfaces of the R2 particles resulting in a slower transport of water into PBs-R2. Generally, the R2 particles had a significantly positive effect on the moisture properties of the PBs, with the coefficients of determination r^2 of the linear correlations "TS or WA = a + b × w/w" ranging from 0.63 to 0.72 (Figures 4 and 5).

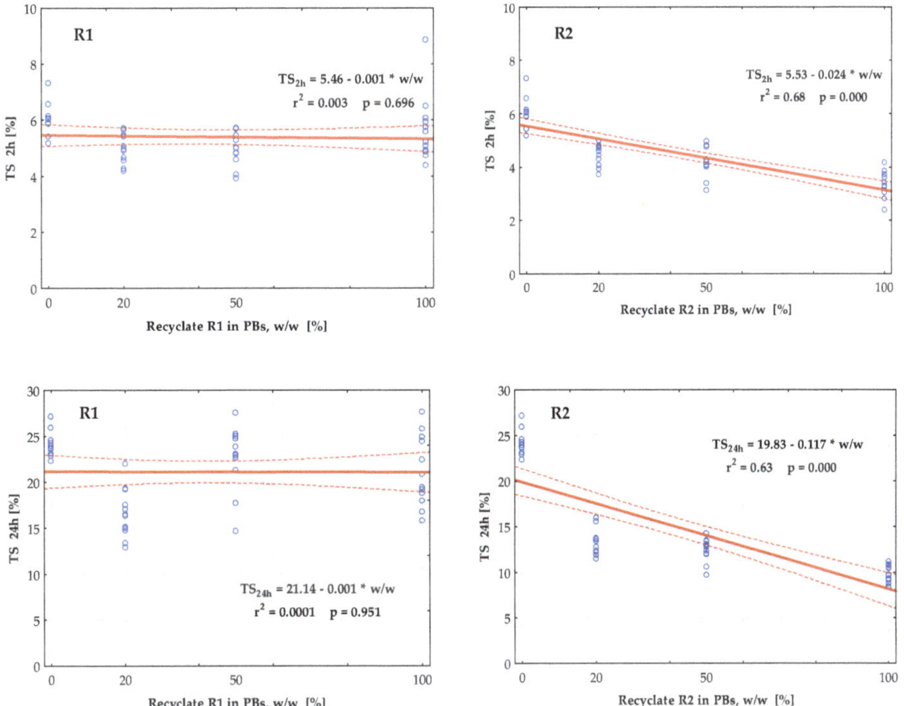

Figure 4. Thickness swelling (TS) after 2 and 24 h of PBs containing different types (R1 or R2) and amounts of wood recyclates.

Similar results were obtained by some other researchers, when in most cases, the moisture properties of PBs based on wood or cellulose waste improved. Azambuja et al. [37] observed the most apparent decrease in thickness swelling (TS) and water absorption (WA) in the case of PBs containing particles from recycled PBs (TS-2h, 5.77%; TS-24h, 15.16%; WA-2h, 17.32%; WA-24h, 44.41%), or particles from the mixture of wood waste "PBs, MDF boards, timber and plywood" (TS-2h, 5.46%; TS-24h, 16.12%; WA-2h, 17.40%; WA-24h, 52.94%). The control PB from pine wood particles had higher values of swelling (TS-2h, 7.49%; TS-24h, 23.16%) and also of water absorption (WA-2h, 24.46%;

WA-24h, 70.47%). However, the PBs containing the recyclate from the plywood showed worse moisture properties (TS-2h, 9.42%; TS-24h, 26.71%; WA-2h, 37.68%; WA-24h, 79.40%) than the control PB.

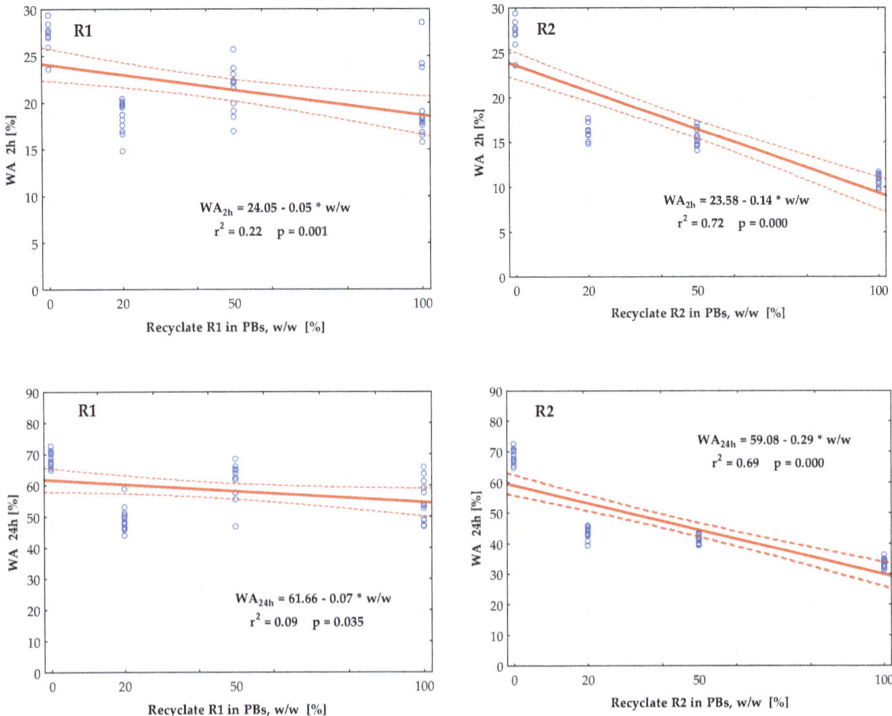

Figure 5. Water absorption (WA) after 2 and 24 h of PBs containing different types (R1 or R2) and amounts of wood recyclates.

Laskowska and Mamiński [36] mentioned that in the case of the recyclates from plywood bonded with UF and PF resin added to the core layer of PBs, the thickness swelling (TS-24h, 8 to 10%) in the boards manufactured from recycled plywood bonded only with UF resin was similar to the swelling in the control PB from wood particles. On the other hand, the thickness swelling of the PBs manufactured based on a waste mixture—i.e., from recycled plywood bonded with PF resin or UF resin and with the recyclate proportions of 25, 50, 75 and 100 wt.%—was 25 to 200% higher compared to the control PB bonded with UF resin. Even when the proportion of the wood particles from the recyclate of plywood bonded with PF resin was 100%, delamination of the PB was observed before testing the swelling.

Weber and Iwakiri [38] studied the moisture properties of PBs manufactured from particles of recycled MDF boards, PBs and plywood. The moisture properties of a PB with a 10 wt.% proportion of UF resin were usually better than the properties of the control PB. The best moisture properties were observed in the case of the PBs manufactured from MDF recyclate (TS-24h, 6.42%; WA-24h, 16.17%). On the contrary, the worst moisture properties were shown for PBs from recycled plywood (TS-24h, 19.80%; WA-24h, 44.33%).

The negative effect of the hydrothermal treatment (HTT) of wood particles on the moisture properties of PBs was mentioned by Lykidis and Grigoriou [17]. The effect of the primary HTT of wood particles, as well as of secondary HTTs, on the thickness swelling and water absorption of PBs was negative (TS-24h, from 37.03 to 59.11%; WA-24h, from 92.47 to 119.61%) in comparison to the control PB (TS-24h, 28.45%; WA-24h, 82.71%).

An interesting experiment was conducted by Nourbakhsh and Ashori [51]. They manufactured PBs with newspaper waste added to poplar (*Populus deltoides*) particles in amounts of 0, 25, 50 and 75 wt.%. The PBs manufactured in this way had TS-24h ranging from 12.1 to 25.9%, also with a dependence on the pressing temperature. For example, when the pressing temperature was 175 °C, the swelling thickness of the PBs with the higher proportion of newspaper was proportionally higher after 24 h: 0 wt.% → TS-24h, 12.1%; 25 wt.% → TS-24h, 14.4%; 50 wt.% → TS-24h, 17.8%; 75 wt.% → TS-24h, 21.6%.

The mechanical properties of the PBs in bending, i.e., the modulus of rupture (MOR) and the modulus of elasticity (MOE), were negatively influenced by both types of recyclates—R1 and R2 (Tables 3 and 4, Figure 6).

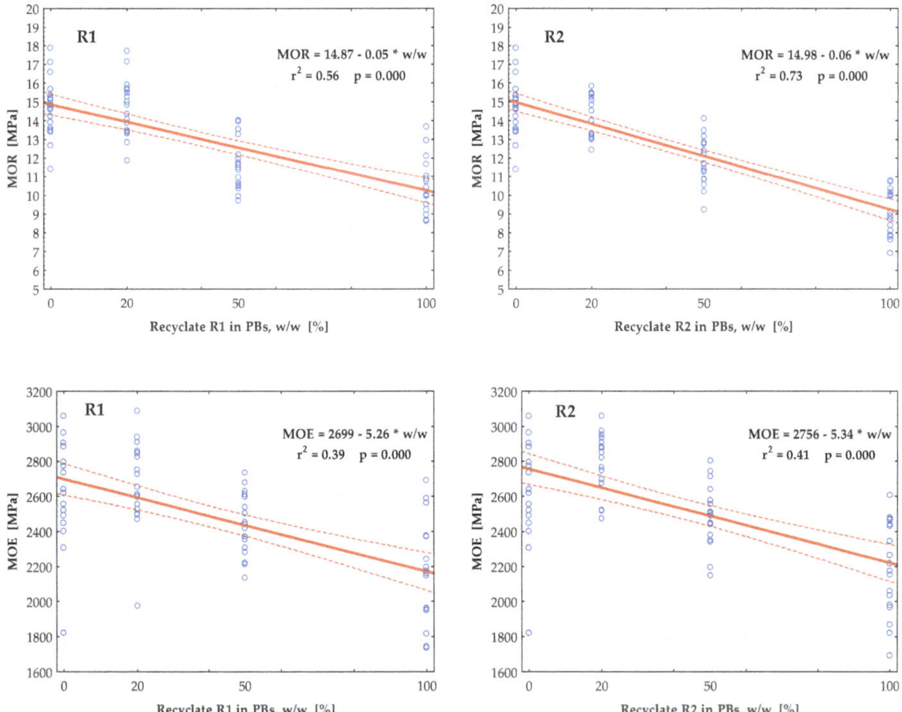

Figure 6. Modulus of rupture (MOR) and modulus of elasticity (MOE) of PBs containing different types (R1 or R2) and amounts of wood recyclates.

The maximal decrease in the MOR, from the 14.7 MPa of PB-C, was determined in the case of using the highest 100% amount of recyclates, i.e., by 28% to 10.6 MPa for PB-R1-100 and by 37% to 9.30 MPa for PB-R2-100. The maximal decrease in the MOE, from the 2637 MPa of PB-C, was determined similarly when the highest 100% amount of recyclates was used, i.e., by 18% to 2155 MPa for PB-R1-100 and by 17% to 2194 MPa for PB-R2-100 (Tables 3 and 4). A significantly negative effect of both wood recyclates, R1 and R2, on the bending properties of the PBs was confirmed by the coefficients of determ ination r² of the linear correlations "MOR or MOE = a + b × *w/w*", ranging from 0.39 for the MOE to 0.73 for the MOR (Figure 6). The mechanical properties (MOR and MOE) of the PBs based on wood recyclates fulfilled the requirements of particleboard type P2 (according to EN 312 [40]: MOR, 11 MPa; MOE, 1600 MPa) except in the MOR for the PB variants with 100% recyclates.

The internal bond (IB) of the control PB-C of 0.82 MPa decreased when there was an increase in the content of R1 or R2 recyclates in the laboratory-prepared PBs, by a maximum of about 33% to 0.55 MPa for PB-R1-100 and also for PB-R2-100 (Tables 3 and 4). A significantly negative effect of the increased amount of wood recyclates R1 and R2 on the IB of the PBs was confirmed by the coefficients of determination r^2 of the linear correlations "IB = a + b × w/w" of 0.65 and 0.70 (Figure 7). Compared to the requirements of the standard EN 312 [40], the PBs based on wood recyclates achieved, in all cases, the IB of 0.35 MPa needed for the type P2, and even 57% higher IB values for the PBs prepared from 100% recyclates.

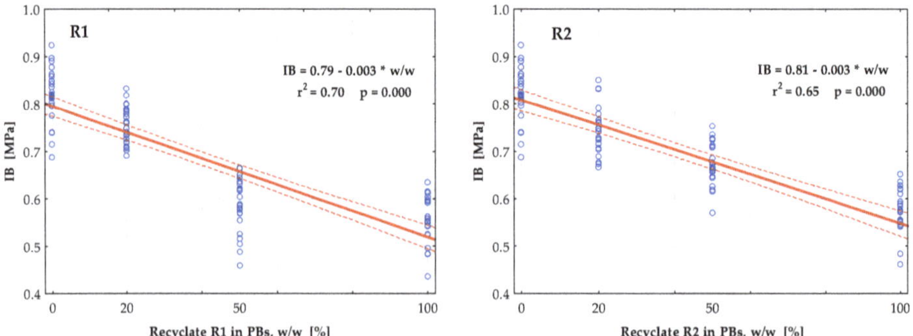

Figure 7. Internal bonds (IB) of PBs containing different types (R1 or R2) and amounts of wood recyclates.

In general, following several experiments—on one hand, those presented in this paper, as well as those conducted by many researchers—it can be stated that negative or, in several cases, even positive changes in the mechanical properties of PBs with wood, lingo-cellulose and cellulose recyclates depend on the type and the portion of particles in the newly manufactured composite board or on the type and portion of the adhesive used. This is confirmed by the results of the following research studies.

Azambuja et al. [37] found out that in many cases, the bending characteristics MOR and MOE of PBs with various types of wood recyclates decreased in comparison to the control PB made from pine particles (MOR, 8.00 MPa; MOE, 1378 MPa). The lowest values of the MOR and MOE, even below the requirements mentioned in the standards, were obtained in the case of PBs manufactured from waste MDF boards (MOR, 4.59 MPa; MOE, 497 MPa) and from PBs (MOR, 6.49 MPa; MOE, 1013 MPa), while low values were also observed in the case of PBs from plywood (MOR, 7.12 MPa; MOE, 1299 MPa). On the contrary, PBs manufactured from the recyclate of timber showed better bending characteristics (MOR, 9.69 MPa; MOE, 1392 MPa) compared to the control PB. This could be caused by the relatively high quality and strength of timber from coniferous trees built in trusses, ceilings and other structural parts over the decades under conditions unsuitable for decay and other deterioration processes in wood [52]. The wood recyclates also affected the IB values of the PBs in different ways—in a range from 0.18 MPa when using the waste from MDF (apparently below the standard level, 0.35 MPa) up to 0.96 MPa when using the recyclate from timber.

Laskowska and Mamiński [36] mentioned that recycled particles from plywood added to the core layer of PBs resulted in a decrease in the MOR by 13 to 91% compared to the control PB. Such a significant decrease in bending strength could be due to the less-narrow shape of wood particles from plywood and their larger area. Therefore, they would not be well bonded together with the UF resin in the newly prepared PBs.

Weber and Iwakiri [38], in the case of PBs based on various wood recyclates, determined a decrease in the MOR and MOE. For example, when using 10% UF resin, the values of the MOR ranged from 5.75 MPa for PBs based on the mixed recyclate from PBs and plywood up to 9.76 MPa for those based on the recyclate from MDF boards. The finding that in the case of the PBs from the recyclate of

MDF, the values of the MOE, 1129 MPa, were the lowest can be considered as interesting—and it was probably caused by the short, thin wood fibers from the MDF boards in prepared PBs. On the contrary, the highest values of the MOE, 1522 MPa, were observed in the case of the mixture of the waste of PBs and plywood. The decrease in IB values, in comparison to the control PB, did not appear in the case of the PBs based on the particles from MDF, PBs, plywood recyclates, and their combinations. The IB values of such prepared PBs ranged from 0.46 MPa when using the recyclate from MDF boards up to 0.56 MPa when using the recyclate from PBs and MDF boards. However, when applying a smaller 6% amount of UF resin to the PBs with wood recyclates, a decrease in all the mechanical properties was more evident. In the case of PBs based on MDF, the value of the IB was only 0.26 MPa.

The strength characteristics (MOR and IB) of the PBs were negatively affected by adding the hydrothermally modified (HTT) wood particles [17]. The value of the MOR in the PBs after the first HTT of the wood particles ranged from 12.11 to 14.24 MPa, and the value of IB, from 0.379 to 0.712 MPa. In the case of the PBs manufactured from the wood particles modified with two subsequent HTTs, the values of the MOR, 9.52 to 13.55 MPa, and of the IB, 0.177 to 0.504 MPa, were significantly lower compared to the values of the control PB (MOR, 17.15 MPa; IB, 0.938 MPa). On the other hand, the MOE values of the PBs containing HTT wood particles even partly increased—for one HTT, from 2198 to 2402 MPa, or for two HTTs, from 2379 to 2581 MPa—compared to the control PB with a MOE of 2137 MPa. This could be due to the higher toughness of wood particles after changes in their molecular structure and, again, obtaining cooled lignin macromolecules in a glassy state in the wood particles.

Nourbakhsh and Ashori [51] found out that when the newspaper/poplar wood waste ratio in PBs was increased, the mechanical properties (MOR, MOE and IB) of the boards were worse. At a pressure temperature of 175 °C, in the case of the PBs with 50% newspaper waste, the MOR was 24.7 MPa, the MOE was 2357 MPa and the IB was 0.7 MPa, or in the case of PBs with 75% newspaper waste, the MOR was 17.8 MPa, the MOE was 1858 MPa and the value of the IB was 0.57 MPa. The reference PB, prepared only from poplar particles, showed better mechanical properties (MOR, 28 MPa; MOE, 2748 MPa; IB, 1.16 MPa).

3.2. Biological Resistance of PBs

The biodegradation of PBs by the brown-rot fungus *Coniophora puteana* was more pronounced if the boards contained the R1 recyclate, and on the contrary, it was suppressed by the addition of the R2 recyclate (Table 5, Figures 8–10). *C. puteana* caused a 11.26% mass loss of the reference PB-C. The PB-R1-100 was characterized by the greatest mass loss, 14.86% (Δm increased by 32%), and the PB-R2-100 had the slightest mass loss, 6.26% (Δm decreased by 44%). This phenomenon is similar to the one mentioned before when evaluating a positive effect of R2 recyclate from faulty PBs on the evidently improved moisture properties of newly prepared PBs. The coefficients of determination r^2 of the linear correlations "Mass loss = a + b × w/w" were not the highest, at 0.24 or 0.33; however, on the basis of them, the above-mentioned tendencies were confirmed with sufficient significance—a more intensive decay of PBs containing the old wood "recyclate R1" and a less intensive decay of PBs containing particles covered with UF resin obtained from the failed PBs, "recyclate R2" (Figure 10).

The increased fungal attack of PBs containing Rl recyclate can be explained by their composition, e.g., (a) the presence of old furniture produced from beech particles over 50–60 years of the 20th century in the "Bukas" PBs from the company Bučina Zvolen, Czechoslovakia; according to EN-350 [53], the decay resistance of beech wood is 5, "non-durable", while that of spruce wood is better, 4, "less durable", (b) the presence of HB and MDF boards containing defibrated fine wood fibers, which are easily accessible to hydrolase and other fungal enzymes. The decreased decay of PBs containing R2 recyclate can be attributed to the higher amount of cured UF resins on the surfaces of wood particles.

Table 5. Biological resistance of PBs containing recyclates R1 or R2—valued on the basis of mass losses (Δm) caused by the brown-rot fungus *Coniophora puteana*.

Recyclate Type in PB	Mass Loss of PB Caused by *C. puteana* (%)			
	Recyclates R1 or R2 in PB *w/w* (%)			
	0	20	50	100
R1	11.26 (2.57)	12.43 (2.14)	12.81 (1.76)	14.86 (3.30)
R2	11.26 (2.57)	7.52 (1.47)	6.69 (2.40)	6.26 (1.61)

Notes: Mean values: mass loss due to *C. puteana* from 6 samples. Standard deviations are in the parentheses. R1 = particles from mixture of several recycled wood products; R2 = particles from recycled faulty PBs.

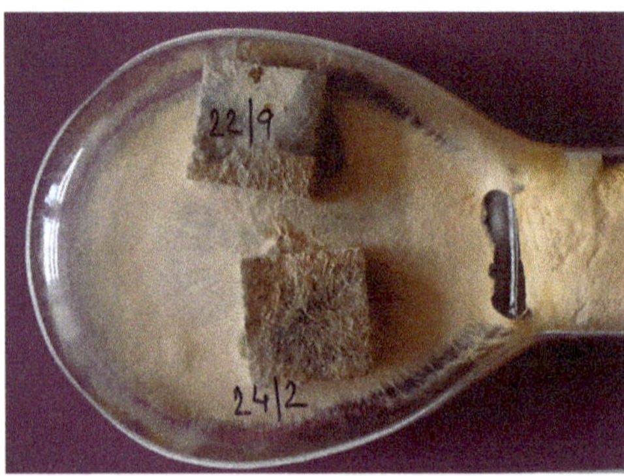

Figure 8. Growth of the surface mycelia of the brown-rot fungus *Coniophora puteana* on the top surfaces of PB-20-R1 (it contains 20% particles from a mixture of recycled wood products R1 and 80% particles from fresh spruce wood) after 16 weeks of the decay test in Kolle flasks.

	0:100	50:50	80:20
R1 PBs containing particles from mixture of recycled wood products			
R2 PBs containing particles from recycled faulty PBs			
Weight relations of fresh spruce wood particles and recycled wood particles R1 or R2 in tested PB			

Figure 9. Growth of the surface mycelia of the brown-rot fungus *Coniophora puteana* on the top surfaces of PBs containing particles from recycled wood after 16 weeks of the decay test.

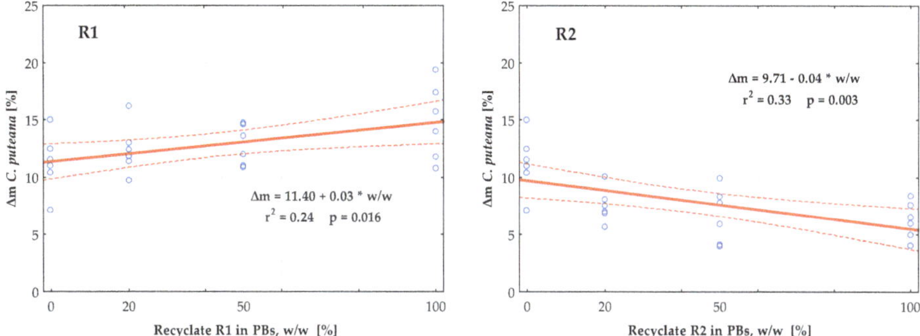

Figure 10. Biological resistance of PBs containing different types (R1 or R2) and amounts of wood recyclates to the brown-rot fungus *Coniophora puteana*, determined by mass losses "Δm" (mycological test according to ENV 12038 [50]).

The interactive effect of the moisture properties and decay resistance on panel composites is also reported by other studies [54–56], in which a higher water absorption by wood particles resulted in a more intensive decay in such composites. The biological resistance of panel products made from waste biomaterials to wood-decaying fungi is also influenced by the type of resin used [55,57] and biocides [58,59] and other additives [60] present in various wood composite types.

4. Conclusions

- The thickness swelling (TS) and the water absorption (WA) of PBs prepared from wood recyclates decreased more apparently when using R2 particles prepared from faulty UF-bonded PBs—with a decrease in TS after 24 h by 59% and in WA after 24 h by 51%. This was probably due to the presence of cured UF resin on the surfaces of such recycled wood particles. Adding the R1 particles from the old wood products did not apparently affect the moisture properties of the PBs.

- The modulus of rupture (MOR) in bending, the modulus of elasticity (MOE) in bending, and the internal bond (IB) of the PBs were negatively affected by a larger amount of R1 and R2 recyclates in them. The most apparent decrease in all the tested mechanical properties was shown for the PBs prepared only from recyclates—containing 100% R1 or R2—i.e., there was a decrease in the MOR from 14.7 to 10.6 or 9.30 MPa, a decrease in the MOE from 2637 to 2155 or 2194 MPa, and a decrease in the IB from 0.82 to 0.55 MPa.

- The biological resistance of PBs to the brown-rot fungus *Coniophora puteana* decreased in the presence of the R1 recyclate, from the old wood products. On the contrary, the R2 recyclate, from the faulty PBs, evidently suppressed the decay activity of *C. puteana* in newly prepared PBs. The R2 recyclate had, in this situation, a similar effect as was determined in the case of the moisture properties of the PBs—probably due to the presence of some portion of cured UF resin on its surfaces.

- Adding the suitable types of wood recyclates to PBs is important in terms of economy and the environment as well. Generally, PBs containing wood recyclates could be used for less or more stressed exposures, considering the used type and amount of recyclate content.

- The three-layered UF resin-bonded PBs based on wood recyclates showed sufficient mechanical properties, complying with EN 312 [40], and fulfilled the requirements for particleboard type P2, except in the MOR for PB variants with 100% recyclates.

Author Contributions: Conceptualization, J.I., Z.V. and L.R.; methodology, J.I., Z.V. and L.R.; software, J.I. and Z.V.; validation, J.I., Z.V. and L.R.; formal analysis, J.I. and L.R.; investigation, J.I., Z.V. and L.R.; resources, J.I., Z.V. and L.R.; data curation, J.I., Z.V. and L.R.; writing—original draft preparation, J.I., Z.V. and L.R.; writing—review and editing, J.I. and L.R.; visualization, J.I. and Z.V.; supervision, L.R.; project administration, J.I. and L.R.; funding acquisition, L.R. All authors have read and agreed to the published version of the manuscript.

Funding: This work was supported by the Slovak Research and Development Agency under the contract no. APVV-17-0583.

Acknowledgments: We would like to thank, for providing some of the materials for this research, the companies Kronospan, s.r.o., Zvolen, Slovakia, and Kronospan CR, spol. s r.o., Jihlava, Czech Republic.

Conflicts of Interest: The authors declare no conflict of interest.

References

1. Kollmann, F.F.P.; Kuenzi, E.W.; Stamm, A.J. *Principles of Wood Science and Technology, II Wood Based Materials*, 1st ed.; Springler: Berlin/Heidelberg, Germany, 1975; p. 703.
2. Barbu, M.C.; Réh, R.; Çavdar, A.D. Non-wood lignocellulosic composites. In *Research Developments in Wood Engineering and Technology*, 1st ed.; Aguilera, A., Davim, J.P., Eds.; IGI Global: Hershey PA, USA, 2014; Volume 8, pp. 281–319. [CrossRef]
3. Nunes, L.; Réh, R.; Barbu, M.C.; Walker, P.; Thomson, A.; Maskell, D.; Knapic, S.; Bajraktari, A.; Greef, J.M.; Brischke, C.; et al. Nonwood bio-based materials. In *Performance of Bio-Based Building Materials*, 1st ed.; Jones, D., Brischke, C., Eds.; Woodhead Publishing: Duxford, UK, 2017; Volume 3, pp. 97–186. ISBN 978-0-08-100982-6.
4. Alma, M.H.; Temiz, A.; Kalaycioglu, H. Wood-based panels from alternative raw material. In *Performance in Use and New Products of Wood Based Composites*, 1st ed.; Fan, M., Ohlmeyer, M., Irle, M., Haelvoet, W., Athanassiadou, E., Rochester, I., Eds.; Brunel University Press: London, UK, 2009; Volume 13, pp. 267–283. ISBN 978-1-902316-75-8.
5. Maloney, T.M. *Modern Particleboard and Dry-Process Fiberboard Manufacturing*, 2nd ed.; Miller Freeman: San Francisco, CA, USA, 1993; pp. 626–669.
6. Deppe, H.J.; Ernst, K. *Taschenbuch der Spanplattentechnik*, 4th ed.; DRW: Leinfelden-Echterdingen, Germany, 2000; p. 552. ISBN 3-87181-349-4.
7. Laskowska, A.; Mamiński, M. The properties of particles produced from waste plywood by shredding in a single-shaft shredder. *Maderas Cienc. Technol.* **2020**, *22*, 197–204. [CrossRef]
8. The European Parliament and the Council. Directive 2008/98/EC of the European Parliament and the Council of 19 November 2008 on waste and repealing certain Directives. *Off. J. Eur. Union* **2008**, *51*, 3–30.
9. Réh, R. *Wood-Based Particle and Fiber Panel Products*, 1st ed.; TU Zvolen: Zvolen, Slovak Republic, 2013; p. 136. ISBN 978-80-228-2498-9.
10. FAOSTAT. Food and Agriculture Organization of the United Nations. Forestry Production and Trade. Available online: http://www.fao.org/faostat/en/#data/FO (accessed on 18 September 2020).
11. Varis, R. (Ed.) Particleboard. In *Wood-Based Panels Industry*, 1st ed.; Finnish Woodworking Engineers Association: Bookwell Oy, Porvoo, Finland, 2018; Volume 4, pp. 197–223. ISBN 978-952-68627-8-1.
12. Kharazipour, A.; Kües, U. Recycling of wood composites and solid wood products. In *Wood Production, Wood Technology, and Biotechnological Impacts*, 1st ed.; Kües, U., Ed.; Universitätsverlag: Göttingen, Germany, 2007; Volume 20, pp. 509–533. ISBN 13: 978-3-940344-11-3.
13. Wan, H.; Wang, X.M.; Barry, A.; Shen, J. Recycling wood composite panels: Characterizing recycled materials. *BioResources* **2014**, *9*, 7554–7565. [CrossRef]
14. Michanickl, A.; Boehme, C. Process for Recovering Chips and Fibers from Residues of Timber-Derived Materials, Old Pieces of Furniture, Production Residues, Waste and Other Timber Containing Materials. U.S. Patent 5804035, 8 September 1998.
15. Roffael, E.; Dix, B.; Behn, C.; Bär, G. Use of UF-bonded recycling particle- and fibreboards in MDF-production. *Eur. J. Wood Prod.* **2010**, *68*, 121–128. [CrossRef]
16. Franke, R.; Roffael, E. Zum recycling von span- und MDF-platten. Teil 1: Über die Hydrolyseresistenz von ausgehärteten Harnstoff-Formaldehydharzen (UF-Harzen) in Span- und mitteldichten Faserplatten (MDF). *Holz als Roh-und Werkstoff* **1998**, *56*, 79–82. [CrossRef]

17. Lykidis, C.; Grigoriou, A. Hydrothermal recycling of waste and performance of the recycled wooden particleboards. *Waste Manag.* **2008**, *28*, 57–63. [CrossRef]

18. Lykidis, C.; Grigoriou, A. Quality characteristics of hydrothermally recycled particleboards using various wood recovery parameters. *Int. Wood Prod. J.* **2011**, *2*, 38–43. [CrossRef]

19. Michanickl, A. *Recycling of Laminated Boards. Asian International Laminates Symposium, Hong Kong, China;* TAPPI Press: Atlanta, Georgia, USA, 1997; pp. 269–274.

20. Ye, K.; Yu, W.; Qu, Y. Influence of water thermal treatment on particle form and recyclability of waste particleboard. *China Wood Ind.* **1998**, *6*, 7–9.

21. Roffael, E. Method for Use of Recycled Lignocellulosic Composite Materials. U.S. Patent Application Publication US 2002/0153107 A1, 24 October 2002.

22. Roffael, E.; Athanassiadou, E.; Mantanis, G. Recycling of particle- and fibreboards using the extruder technique. In Proceedings of the 2nd Conference on Environmental Protection in the Wood Industry (Umweltschutz in der Holzwerkstoffindustrie), Goettingen, Germany, 21–22 March 2002; Georg-August-Universität Göttingen: Goettingen, Germany; pp. 56–65.

23. Mantanis, G.; Athanassiadou, E.; Nakos, P.; Coutinho, A. A new process for recycling waste fiberboards. In Proceedings of the 38th International Wood Composites Symposium, Washington, DC, USA, 5–8 April 2004; Pullman, Wash.: Washington, DC, USA, 2004; pp. 119–122.

24. Czarnecki, R.; Dziurka, D.; Łęcka, J. The use of recycled boards as the substitute for particles in the centre layer of particleboards. *Electron. J. Pol. Agric. Univ.* **2003**, *6*. Available online: http://www.ejpau.media.pl/volume6/issue2/wood/art-01.html (accessed on 16 September 2020).

25. Rowell, R.M.; Spelter, H.; Arola, R.A.; Davis, P.; Friberg, T.; Hemingway, R.W.; Rials, T.; Luneke, D.; Narayan, R.; Simonsen, J.; et al. Opportunities for composites from recycled wastewood-based resources: A problem analysis and research plan. *For. Prod. J.* **1993**, *43*, 55–63.

26. Grigoriou, A. The ecological importance of wood products. *Sci. Ann. Dep. For. Nat. Environ.* **1996**, *39*, 703–714.

27. Ihnát, V.; Lübke, H.; Russ, A.; Borůvka, V. Waste agglomerated wood materials as a secondary raw material for chipboards and fibreboards Part I. Preparation and characterization of wood chips in terms of their reuse. *Wood Res.* **2017**, *62*, 45–56.

28. Ihnát, V.; Lübke, H.; Balberčák, J.; Kuňa, V. Size reduction downcycling of waste wood. Review. *Wood Res.* **2020**, *65*, 205–220. [CrossRef]

29. Werner, F.; Althaus, H.J.; Richter, K. Post-consumer wood in environmental decision-support tools. *Schweiz. Z. Forstwes.* **2002**, *153*, 97–106. [CrossRef]

30. Wilson, J.B. Life-cycle inventory of particleboard in terms of resources, emissions, energy and carbon. *Wood Fiber Sci.* **2010**, *42*, 90–106.

31. Parobek, J.; Paluš, H.; Kaputa, V.; Šupín, M. Analysis of wood flows in Slovakia. *BioResoureces* **2014**, *9*, 6453–6462. [CrossRef]

32. Kutnar, A. Environmental Use of Wood Resources. In *Environmental Impacts of Traditional and Innovative Forest-Based Bioproducts. Environmental Footprints and Eco-design of Products and Processes;* Kutnar, A., Muthu, S.S., Eds.; Springer: Singapore, 2016; pp. 1–18. ISBN 978-981-10-0653-1.

33. Ormondroyd, G.A.; Spear, M.J.; Skinner, C. The opportunities and challenges for re-use and recycling of timber and wood products within the construction sector. In *Environmental Impacts of Traditional and Innovative Forest-Based Bioproducts. Environmental Footprints and Eco-design of Products and Processes;* Kutnar, A., Muthu, S.S., Eds.; Springer: Singapore, 2016; pp. 45–104. ISBN 978-981-10-0653-1.

34. Gaff, M.; Trgala, K.; Adamová, T. *Environmental Benefits of Using Recycled Wood in the Production of Wood-Based Panels,* 1st ed.; CZU: Prag, Czech Republic, 2018; pp. 1–51. ISBN 978-80-213-2852-5.

35. Irle, M.; Barbu, M.C. Wood-based panel technology. In *Wood-Based Panels—An Introduction for Specialists,* 1st ed.; Thoemen, H., Irle, M., Sernek, M., Eds.; Brunel University Press: London, UK, 2010; Volume 1, pp. 1–90. ISBN 978-1-902316-82-6.

36. Laskowska, A.; Mamiński, M. Properties of particleboard produced from post-industrial UF- and PF-bonded plywood. *Eur. J. Wood Prod.* **2018**, *76*, 427–435. [CrossRef]

37. Azambuja, R.R.; Castro, V.G.; Trianoski, R.; Iwakiri, S. Recycling wood waste from construction and demolition to produce particleboards. *Maderas Cienc. Technol.* **2018**, *20*, 681–690. [CrossRef]

38. Weber, C.; Iwakiri, S. Utilization of waste of plywood, MDF, and MDP for the production of particleboards. *Ciência Florest.* **2015**, *25*, 405–413. [CrossRef]

39. Hameed, M.; Rönnols, E.; Bramryd, T. Particleboard based on wood waste material bonded by leftover cakes of rape oil. Part 1: The mechanical and physical properties of particleboard. *Holztechnologie* **2019**, *6*, 31–39.

40. EN 312. *Particleboards—Specifications*; European Committee for Standardization: Brussels, Belgium, 2010.

41. EN 827. *Adhesives—Determination of Conventional Solids Content and Constant Mass Solids Content*; European Committee for Standardization: Brussels, Belgium, 2005.

42. EN ISO 2431. *Paints and Varnishes—Determination of Flow Time by Use of Flow Cups (ISO 2431:2019)*; European Committee for Standardization: Brussels, Belgium, 2019.

43. EN 1245. *Adhesives—Determination of pH*; European Committee for Standardization: Brussels, Belgium, 2011.

44. EN 323. *Wood-Based Panels—Determination of Density*; European Committee for Standardization: Brussels, Belgium, 1993.

45. EN 322. *Wood-Based Panels—Determination of Moisture Content*; European Committee for Standardization: Brussels, Belgium, 1993.

46. EN 317. *Particleboards and Fibreboards—Determination of Swelling in Thickness after Immersion in Water*; European Committee for Standardization: Brussels, Belgium, 1993.

47. STN 490164. *Particle Boards—Determination of Water Absorption*; Slovak Office of Standards, Metrology and Testing: Bratislava, Slovak Republic, 1980.

48. EN 310. *Wood-Based Panels—Determination of Modulus of Elasticity in Bending and of Bending Strength*; European Committee for Standardization: Brussels, Belgium, 1993.

49. EN 319. *Particleboards and Fibreboards—Determination of Tensile Strength Perpendicular to the Plane of the Board*; European Committee for Standardization: Brussels, Belgium, 1993.

50. ENV 12038. *Durability of Wood and Wood-Based Products—Wood-Based Panels—Method of Test for Determining the Resistance against Wood-Destroying Basidiomycetes*; European Committee for Standardization: Brussels, Belgium, 2002.

51. Nourbakhsh, A.; Ashori, A. Particleboard made from waste paper treated with maleic anhydride. *Waste Manag. Res.* **2010**, *28*, 51–55. [CrossRef] [PubMed]

52. Reinprecht, L. *Wood Deterioration, Protection and Maintenance*, 1st ed.; John Wiley & Sons Ltd.: Chichester, UK, 2016; p. 357. ISBN 978-1-119-10653-1.

53. EN 350. *Durability of Wood and Wood-Based Products—Testing and Classification of the Durability to Biological Agents of Wood and Wood-Based Materials*; European Committee for Standardization: Brussels, Belgium, 2016.

54. Karimi, A.N.; Tajvidi, M.; Pourabbasi, S. Effect of compatibilizer on the natural durability of wood flour/high density polyethylene composites against rainbow fungus (Coriolus versicolor). *Polym. Compos.* **2007**, *28*, 273–277. [CrossRef]

55. Shalbafan, A.; Benthien, J.T.; Lerche, H. Biological characterization of panels manufactured from recycled particleboards using different adhesives. *BioResources* **2016**, *11*, 4935–4946. [CrossRef]

56. Cavdar, A.D.; Tomak, E.D.; Mengeloglu, F. Long-term leaching effect on decay resistance of wood-plastic composites treated with boron compounds. *J. Polym. Environ.* **2018**, *26*, 756–764. [CrossRef]

57. Sen, S.; Ayrilmis, N.; Candan, Z. Fungicide and insecticide properties of cardboard panels made from used beverage carton with veneer overlay. *Afr. J. Agric. Res.* **2010**, *5*, 159–165.

58. Reinprecht, L.; Vidholdová, Z.; Iždinský, J. Particleboards prepared with addition of copper sulphate. Part 1: Biological resistance. *Acta Fac. Xylologiae Zvolen* **2017**, *59*, 53–60. [CrossRef]

59. Reinprecht, L.; Iždinský, J.; Vidholdová, Z. Biological resistance and application properties of particleboards containing nano-zinc oxide. *Adv. Mater. Sci. Eng.* **2018**, 1–8. [CrossRef]

60. Pizzi, A. Wood products and green chemistry. *Ann. For. Sci.* **2016**, *73*, 185–203. [CrossRef]

Publisher's Note: MDPI stays neutral with regard to jurisdictional claims in published maps and institutional affiliations.

Article

Particleboards from Recycled Thermally Modified Wood

Ján Iždinský, Zuzana Vidholdová and Ladislav Reinprecht

Department of Wood Technology, Faculty of Wood Sciences and Technology, Technical University in Zvolen, T. G. Masaryka 24, 960 01 Zvolen, Slovakia; zuzana.vidholdova@tuzvo.sk (Z.V.); reinprecht@tuzvo.sk (L.R.)
* Correspondence: jan.izdinsky@tuzvo.sk

Abstract: In recent years, the production and consumption of thermally modified wood (TMW) has been increasing. Offcuts and other waste generated during TMWs processing into products, as well as already disposed products based on TMWs can be an input recycled raw material for production of particleboards (PBs). In a laboratory, 16 mm thick 3-layer PBs bonded with urea-formaldehyde (UF) resin were produced at 5.8 MPa, 240 °C and 8 s pressing factor. In PBs, the particles from fresh spruce wood and mixed particles from offcuts of pine, beech, and ash TMWs were combined in weight ratios of 100:0, 80:20, 50:50 and 0:100. Thickness swelling (TS) and water absorption (WA) of PBs decreased with increased portion of TMW particles, i.e., TS after 24 h maximally about 72.3% and WA after 24 h maximally about 64%. However, mechanical properties of PBs worsened proportionally with a higher content of recycled TMW—apparently, the modulus of rupture (MOR) up to 55.5% and internal bond (IB) up to 46.2%, while negative effect of TMW particles on the modulus of elasticity (MOE) was milder. Decay resistance of PBs to the brown-rot fungus *Serpula lacrymans* (Schumacher ex Fries) S.F. Gray increased if they contained TMW particles, maximally about 45%, while the mould resistance of PBs containing TMW particles improved only in the first days of test. In summary, the recycled TMW particles can improve the decay and water resistance of PBs exposed to higher humidity environment. However, worsening of their mechanical properties could appear, as well.

Keywords: particleboards; thermally modified wood; recycled wood; physical and mechanical properties; decay and mould resistance; FTIR

Citation: Iždinský, J.; Vidholdová, Z.; Reinprecht, L. Particleboards from Recycled Thermally Modified Wood. *Forests* **2021**, *12*, 1462. https://doi.org/10.3390/f12111462

Academic Editor: Luis García Esteban

Received: 8 October 2021
Accepted: 24 October 2021
Published: 27 October 2021

Publisher's Note: MDPI stays neutral with regard to jurisdictional claims in published maps and institutional affiliations.

1. Introduction

For the production of particleboards (PBs), nowadays, the trend is to use non-wood lignocellulosic raw materials and especially recycled wood, except of the various traditional forms of fresh natural soft- and hardwood species, as are slabs, edge trimmings, chips, or sawdust. Recycled wood is the subject of much research, especially in terms of the different subtle composition [1]—old wood, e.g., from building constructions, furniture and packaging, and also residual wood from industry. Thermally modified wood (TMW) can be one such residual lignocellulosic material from industrial production.

Wood modification (chemical, thermal, biological) represents an assortment of innovative processes adopted to improve the physical, mechanical, or aesthetic properties of sawn timber, veneer or wood particles used in the production of wood composites. The modified wood should be nontoxic under service conditions. Furthermore, there should be no release of any toxic substances during service, or at the end of life following disposal or recycling the modified wood [2,3]. Products from modified wood do not cause environmental hazards greater than those associated with the disposal of unmodified wood [3]. It means that recycled modified wood is a suitable input raw material for new wooden products, including PBs.

Thermal modification of wood is preferentially aimed at increasing its biological resistance and dimensional stability during climate changes, whereby this wood is characterised with brown-darker colours in its all volume [2,4–7]. TMWs have higher resistance to brown-rot fungi compared to the white-rot fungi, e.g., Sivrikaya et al. [8] showed that mass loss of TM spruce wood due to its 8-weeks attack by the brown-rot fungus *Coniophora puteana* was only 1.4%. However, TMW is more brittle having increased susceptibility to cracking, and its strength properties decrease usually from 10% to 20% [2,7,9].

Several reviews on the topic of thermal modification of wood from different perspectives were performed [2,7,10–14], including its thermal pretreatments with the aim to improve properties of produced wooden composites [15].

Thermal modification of wood dates back as far as 1915. Currently, the TMW is industrially produced, and its use is increasing. The global production volume of modified wood in the world is estimated to be 1,608,000 m^3 per year, which is dominated by thermal modification to be 1,110,000 m^3 per year. Today in Europe were developed commercial processes, e.g., in France (Retification), Finland (ThermoWood), Germany (OHT, Menz Holz), Austria (Thermoholz), Suisse (Intemporis), the Netherlands (Plato Wood), and Estonia (Thermory). It allows the growth of TMW production in Europe, especially when used in exterior cladding, decking and joinery applications—so now it exceeds 695,000 m^3 [3,12,16–18]. In Slovakia, there are few small producers, the highest TECHNI-PAL (Banská Bystrica) with annual production of TMW less than 500 m^3 per year. Import of TMW from other European countries to Slovakia is app. 1000 m^3 per year. Along with increased trend of TMW production in the world, there is and will also be more and more waste of TMW. In the future, post-consumer of TMW as a raw material for further processing and not necessarily for energy purposes, is expected to emerge [4,19]—e.g., for PBs, for which world production is now by FAOSTAT approximately 102 million m^3 [20].

Pelaez-Samaniego et al. [15] summarized the use of TMW for production of wood composites. The properties of PBs manufactured with content of TMW are by more authors [19,21–28] as follows: better dimensional stability; reduced water absorption; improved MOE; similar MOR (although some authors showed reduction of MOR); reduced internal bond. Oriented strand boards (OSB) with content of TMW obtained similar changes in properties [29–37]: e.g., better dimensional stability; better resistance to decay; at which wood species and thermal pretreatment conditions differently affected (reduced or improved) their mechanical properties.

The aim of this work was to study the effect of the thermally modified particles (from industrial offcuts of TMWs) on the selected physical, mechanical and biological properties of PBs prepared in a laboratory from TMW particles and particles of freshly cut spruce logs.

2. Materials and Methods

2.1. Materials

2.1.1. Wood Particles

The spruce wood particles were made from fresh spruce logs (FSL). Used logs were firstly chipped in the company Kronospan s.r.o. Zvolen, Slovakia, and then the chips milled on particles in the grinding SU1 impact cross mill (TMS, Pardubice, Czech Republic) in the laboratories of the Technical University in Zvolen (TUZVO), Slovakia (Figure 1). Prepared spruce particles were used for the production of reference particleboards (PBs), and also for the production of PBs containing different proportions of TMW particles.

The TMW particles were prepared from waste offcuts of three TMWs—pine, beech and ash—with had a constant dimension of 25 mm × 150 mm × 800 mm. Offcuts of TMWs were obtained from the timber plant TECHNI-PAL, Banská Bystrica, Slovakia. The industrial production of TMWs themselves was performed by a classic technology, i.e., thermal treatment in air at a temperature of 195 °C for a period of 3 h. The spruce wood and TMWs had the following compression strength parallel with grains (averages determined from 15 samples 20 mm × 20 mm × 30 mm at moisture content of 4.7–6.2%): spruce 42.3 MPa; pine-TMW 52.4 MPa; beech-TMW 85.3 MPa; ash-TMW 78.6 MPa—i.e., Average-TMWs 72.1 MPa.

The mixture of TMW particles used for the experiment contained one-third of pine, beech and ash TMWs. In the laboratories of TUZVO, the TMW offcuts were firstly chipped using the 230H drum mower (Klöckner KG, Hirtscheid—Erbach, Westerwald, Germany), and subsequently milled on particles using the grinding SU1 impact cross mill (TMS, Pardubice, Czech Republic) (Figure 1).

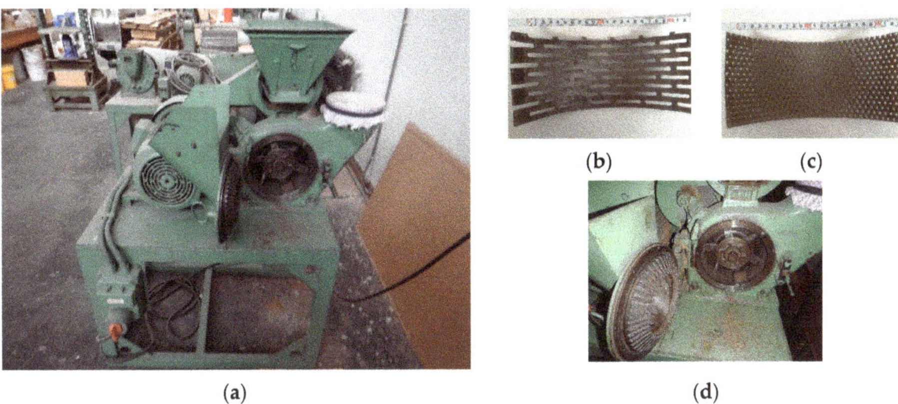

Figure 1. Milling of wood chips using the SU1 impact cross mill (TMS, Pardubice, Czech Republic): (**a**) view of the equipment with grinding segments defining the particle size; (**b**) grate for central particles; (**c**) grate for surface particles; (**d**) a detailed view of the grinding device.

The spruce wood and TMW particles (Figure 2) for the core layer of PBs had dimensions from 0.25 mm to 4.0 mm, and for the surface layers from 0.125 mm to 1.0 mm. The particles for the core layer were dried to a moisture content of 2% and for the surface layers to a moisture content of 4%.

Figure 2. Particles for particleboards (PBs): (**a**) The spruce (**I.**) and thermally modified wood (TMW) particles of (**II.**) pine; (**III.**) beech; (**IV.**) ash used in surface layers (SL) and core layer (CL); (**b**) Size characteristics of the spruce wood and mixture of TMW particles—percentages of individual fraction for surface layers (SL) and core layer (CL).

Particles with a dimension of ≤ 0.25 mm were subjected to FTIR spectra analysis performed on a Nicolet iS10 spectrometer equipped with Smart iTR ATR accessory using diamond crystal. For all particle types, 4 spectral measurements were performed in the range from 4000 cm^{-1} to 650 cm^{-1} with a resolution of 4 cm^{-1}. Measured spectra were baseline corrected and analyzed in absorbance mode by OMNIC 8.0 software (Table 1).

Table 1. Intensity of FTIR spectra (normalized at 898 cm^{-1}) for particles from the fresh spruce logs (FSL) and thermally modified woods (TMWs)—pine, beech and ash.

FTIR (cm^{-1})	Spruce-FSL	Pine-TMW	Beech-TMW	Ash-TMW	Average-TMWs
1274	1.64	0.33	0.25	0.31	0.30
1334	0.15	0.16	0.13	0.19	0.16
1372	1.11	1.10	1.35	1.31	1.25
1430	1.06	0.97	1.31	1.54	1.27
1510	2.42	2.28	1.92	3.00	2.40
1600	0.75	1.09	1.91	2.31	1.77
1653	0.17	0.06	0.08	0.05	0.06
1730	1.09	1.30	2.91	2.46	2.22
2900	1.60	1.61	1.46	1.64	1.57
TCI = 1372/2900	0.69	0.68	0.92	0.80	0.80
LOI = 1430/898	1.06	0.97	1.31	1.54	1.27

2.1.2. Resin and Additives

The UF resin KRONORES CB 4005 D was added into the surface layers of PBs in amount of 11%, and the UF resin KRONORES CB 1637 D was added into the core layer of PBs in amount of 7%. Molar ratio of UF resins was 1.2 (U:F). Other physical-chemical characteristics of used UF resins were mentioned in the work Iždinský et al. [38]. Required curing of UF resins was achieved by the hardener ammonium nitrate, used as 57 wt.% water solution. Hardener was applied in amount of 2% or 4% to the dry mass of UF resins used for the surface or core particles. Paraffin, used as 35 wt.% water emulsion, was applied on the surface and core particles in amount of 0.6% and 0.7%, respectively.

2.2. Particleboard Preparation

The 3-layer PBs with the dimension of 400 mm × 300 mm × 16 mm and with the density of 650 kg·m^{-3} ± 10 kg·m^{-3} were produced under laboratory conditions. UF resin with the hardener was applied on the conditioned wood particles in laboratory rotary mixing device (TU Zvolen, Slovakia). Wood particles mixed with the adhesive had moisture content from 9.1% to 10.4% (for the surface layers) and from 6.3% to 7.1% (for the core layer). Surface/core particles ratio was 35:65. The particle mats were layered manually in wooden forms. Each particle mat was cold pre-pressed at a laboratory temperature, at a pressure of 1 MPa, and then it was pressed in the CBJ 100-11 laboratory press (TOS, Rakovník, Czech Republic) in accordance with the pressing diagram (Figure 3), i.e., at a maximum temperature of 240 °C, a maximum pressing pressure of 5.75 MPa, and a pressing factor of 8 s/mm. In total, 24 PBs were manufactured, i.e., 6 from each type (Table 2).

Table 2. Individual types of manufactured particleboards (PBs).

Variant of PB	Thermally Modified Wood Amount in PB, *w/w* (%)	Number of Produced Boards	Board Type
PB-C:			
100% particles of spruce wood	0	6	C
PB-TMW:			
20%, 50% or 100% particles from	20	6	TMW-20
mixture of thermally modified			
woods (TMWs),	50	6	TMW-50
combined with 80%, 50% or 0%	100	6	TMW-100
particles of spruce wood			

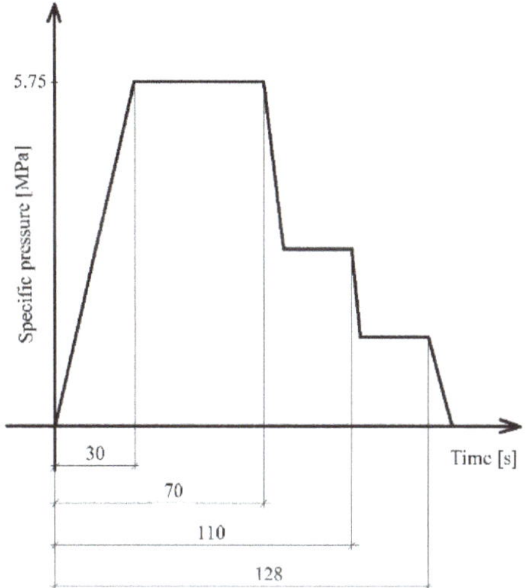

Figure 3. Standard three stage pressing diagram in the manufacturing of particleboards.

2.3. Physical and Mechanical Properties of PBs

Physical and mechanical properties of PBs were determined by the European (EN) and Slovak (STN) standards: the density by EN 323 [39]; the moisture content by EN 322 [40]; the thickness swelling (TS) and the water absorption (WA) after 2 and 24 h by EN 317 [41] and STN 490164 [42]; the modulus of rupture (MOR) in bending and the modulus of elasticity (MOE) in bending by EN 310 [43]; and the internal bond (IB)—the tensile strength perpendicular to the plane of PB—by EN 319 [44]. Samples were prepared from the 4-weeks air-conditioned PBs (Figure 4). Their mechanical properties were determined with the universal machine TiraTest 2200 (VEB TIW, Rauenstein, Germany).

Classification of PBs was performed in accordance with the European standard EN 312 [45], taking into account the requirements of the PB (type P2) with the thickness ranging between 13 mm and 20 mm.

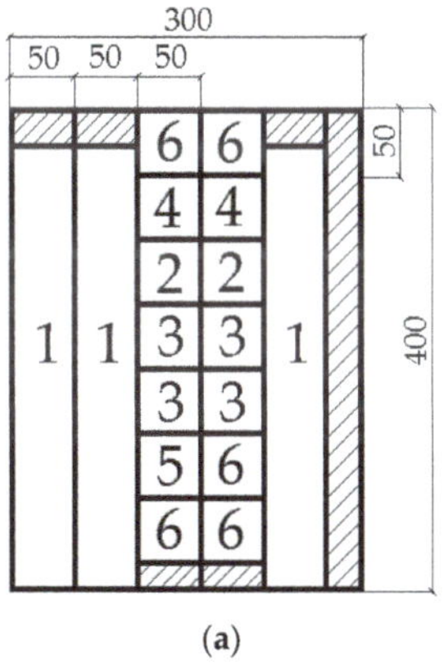

(a)

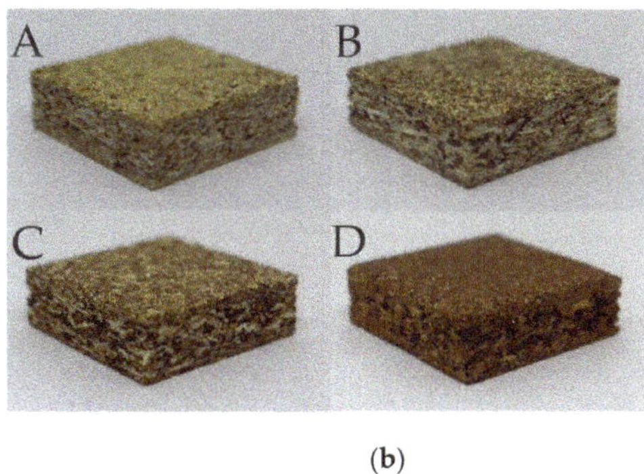

(b)

Figure 4. Scheme of samples preparation from the particleboard (PB) (**a**), i.e., 1—MOR and MOE by 3-point bending test [43], 2—TS and WA after 2 and 24 h [41,42], 3—IB [44], density [39], and moisture content [40], 4—decay resistance to the fungus *Serpula lacrymans* [46], 5—mould resistance [47], 6—spare samples; Display of PB-samples 50 mm × 50 mm × 16 mm with different amount of recycled TMW w/w (%) (**b**), i.e., (**A**) 0%, (**B**) 20%, (**C**) 50%, and (**D**) 100%.

2.4. Decay Resistance of PBs

The decay resistance of PBs to in the interiors very dangerous brown-rot fungus *Serpula lacrymans* (Schumacher ex Fries) S.F. Gray/*S. lacrymans* (Wulfen) J. Schröt.—by IndexFungorum/, strain BAM 87 (Bundesanstalt für Materialforshung und -prüfung, Berlin) was tested on samples 50 mm × 50 mm × 16 mm. Before the mycological test, performed by ENV 12038 [46]: (a) the edges of samples (50 mm × 16 mm) were sealed in the amount of 200 ± 10 g·m^{-2} with epoxy resin CHS-Epoxy 1200 mixed with the hardener P11 in weight ratio 11:1 (Stachema, Mělník, Czech Republic); (b) the samples were oven dried and sterilized at rising temperature from 60 °C to 103 ± 2 °C for 10 h, cooled in desiccators, and weighed (m_0); (c) all surfaces of samples were repeatedly sterilized twice for 30 min with a UV light radiator; (d) the samples were soaked in distilled water for 240 min to achieve the moisture content of 25% to 30%; (e) the samples were placed into 1 L Kolle-flasks on a top of the stainless steel grids with the fungal mycelium inoculant grown on the agar-malt soil (HiMedia Laboratories Pvt. Ltd., Mumbai, India)—into each flask one sample of PB containing recycled TMW and one sample of reference PB.

Kolle-flasks with samples were incubated for 16 weeks at a temperature of 22 ± 2 °C and a relative humidity of 75% to 80%. After the mycological test, the samples of PBs—after depriving the fungal mycelia from their surfaces—were underwent to a gradual drying process to the constant weight ($m_{0/Fungally-Attacked/}$), i.e., 100 h at 20 °C, 1 h at 60 °C, 1 h at 80 °C, 8 h at 103 ± 2 °C, and final cooling in desiccators. Their mass losses (Δm) in percentage were calculated by Equation (1), and their moisture contents (w) in percentages by Equation (2).

$$\Delta m = \frac{m_0 - m_{0/Fungally-Attacked/}}{m_0}\, 100 \;(\%) \tag{1}$$

$$w = \frac{m_{w/Fungally-Attacked/} - m_{0/Fungally-Attacked/}}{m_{0/Fungally-Attacked/}}\, 100 \;(\%) \tag{2}$$

2.5. Mould Resistance of PBs

The mould resistance of PBs against the mixture of microscopic fungi (moulds) often active in interiors (*Aspergillus versicolor* BAM 8, *Aspergillus niger* BAM 122, *Penicillium purpurogenum* BAM 24, *Stachybotrys chartarum* BAM 32 and *Rhodotorula mucilaginosa* BAM 571) was tested according to a partly modified EN 15457 [47]—it means the samples from PBs had other size, i.e., 50 mm × 50 mm × 16 mm, and their sterilization was performed by other method, i.e., twice for 30 min by UV light radiator. Air-conditioned samples were placed into Petri dishes with a diameter of 180 mm (two pieces in one dish) on the Czapek-Dox agar soil (HiMedia Laboratories Pvt. Ltd., Mumbai, India), and then were inoculated with water suspension of mould spores. Incubation lasted 28 days at a temperature of 24 ± 2 °C and relative humidity of 90–95%.

Mould growth activity (MGA) on the top surfaces of PBs was assessed on the 7th, 14th, 21st day (and also on the 28th day—in addition to EN 15457 [47]) after inoculation by a scale from 0 to 4, using the following criteria: 0 = no mould on the surface; 1 = mould up to 10%; 2 = mould up to 30%; 3 = mould up to 50%; 4 = mould more than 50% on the surface.

2.6. Statistical Analyses

The statistical software STATISTICA 12 (StatSoft, Inc., Tulsa, OK, USA) was used to analyze the gathered data. Descriptive statistics deals with basic statistical characteristics of studied properties—arithmetic mean and standard deviation. The simple linear correlation analysis, determining also the coefficient of determination, was used as method of inductive statistic to evaluate the measured data.

3. Results and Discussion

3.1. Physical and Mechanical Properties of PBs

The basic physical and mechanical properties of PBs manufactured in the laboratories of TUZVO are presented in Table 3. The effect of recycled TMWs on the properties of PBs was analyzed by the linear correlations and the coefficients of determination r^2 in Figures 5–9.

Table 3. Physical and mechanical properties of the reference/control particleboard (PB-C) and the particleboards (PBs) containing particles from a mixture of recycled thermally modified wood offcuts (PB-TMW).

Property of PB		Thermally Modified Wood (TMW) in PB *w/w* (%)			
		0	20	50	100
Density	(kg·m^{-3})	654 (15.7)	659 (23.6)	657 (21.7)	653 (23.6)
Thickness swelling (TS) after 2 h	(%)	6.01 (0.53)	3.78 (0.47)	3.00 (0.27)	1.97 (0.45)
Thickness swelling (TS) after 24 h	(%)	23.73 (1.38)	15.06 (0.61)	10.73 (1.12)	6.58 (0.59)
Water absorption (WA) after 2 h	(%)	27.40 (2.04)	16.34 (0.91)	12.16 (0.81)	6.38 (0.42)
Water absorption (WA) after 24 h	(%)	68.18 (2.31)	45.28 (2.04)	37.32 (1.51)	24.55 (2.24)
Internal bond (IB)	(MPa)	0.78 (0.05)	0.71 (0.04)	0.61 (0.06)	0.42 (0.06)
Modulus of rupture (MOR)	(MPa)	14.6 (1.56)	13.1 (1.45)	10.3 (0.96)	6.5 (0.74)
Modulus of elasticity (MOE)	(MPa)	2611 (285)	2649 (390)	2189 (443)	2225 (253)

Notes: Mean values of density are from 42 samples, TS from 12 samples, WA from 12 samples, IB from 24 samples, MOR and MOE from 18 samples. Standard deviations are in the parentheses.

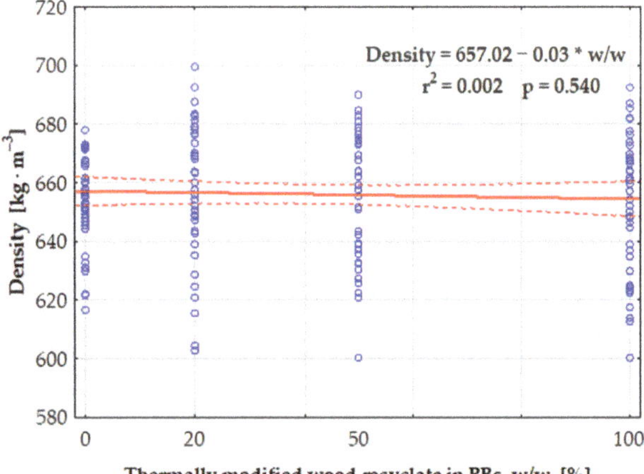

Figure 5. Density of PBs containing different amount of thermally modified wood recyclates.

The average density of PBs ranged in a very narrow interval between 653 kg·m^{-3} and 659 kg·m^{-3}, i.e., without any apparent influence of the different mutual weight portion of particles from the FSL and recycled TMWs (Table 3). In the linear correlation, it confirmed the r^2 value lower than 0.002 (Figure 5). Following it means that the tested moisture, mechanical and biological properties of PBs could not be affected by the density factor.

However, in theory, the searched properties of PBs could be affected by the structure and properties of used wood species. In this study the following analyses were performed: (1) FTIR spectroscopy of wood particles for identification some molecular characteristics of the lignin-polysaccharide matrix in cells walls (Table 1); (2) Compression strength of wood parallel with grains (Section 2.1.1). An attempt to explain the differences in selected properties of PBs by the FTIR and the strength of the types of wood used is presented at the end of the discussion.

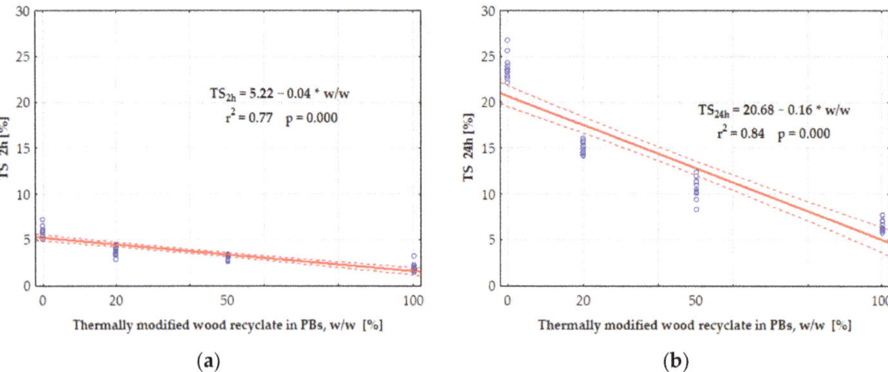

(a) **(b)**

Figure 6. Thickness swelling (TS) after 2 h (**a**) and 24 h (**b**) of PBs containing different amount of thermally modified wood recyclates.

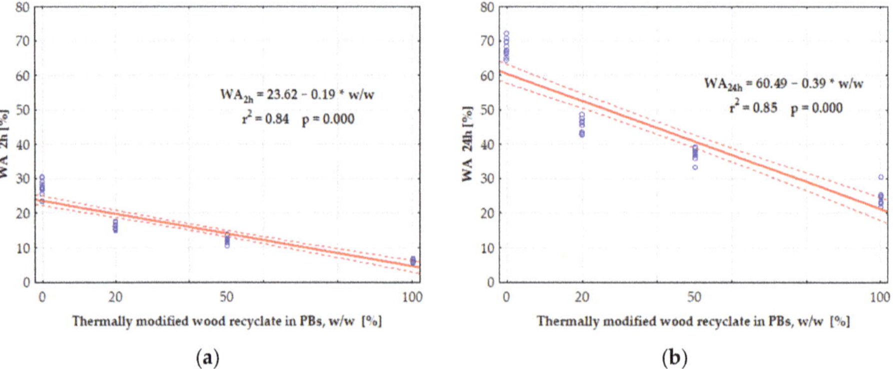

(a) **(b)**

Figure 7. Water absorption (WA) after 2 h (**a**) and 24 h (**b**) of PBs containing different amount of thermally modified wood recyclates.

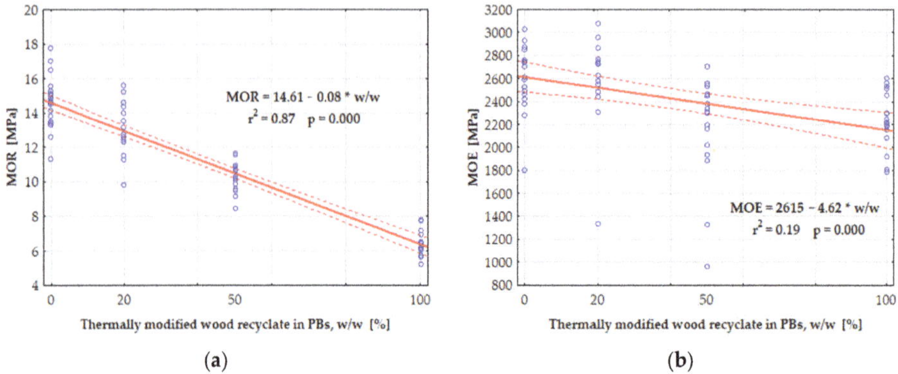

(a) **(b)**

Figure 8. Modulus of rupture (MOR) (**a**) and modulus of elasticity (MOE) (**b**) of PBs containing different amount of thermally modified wood recyclates.

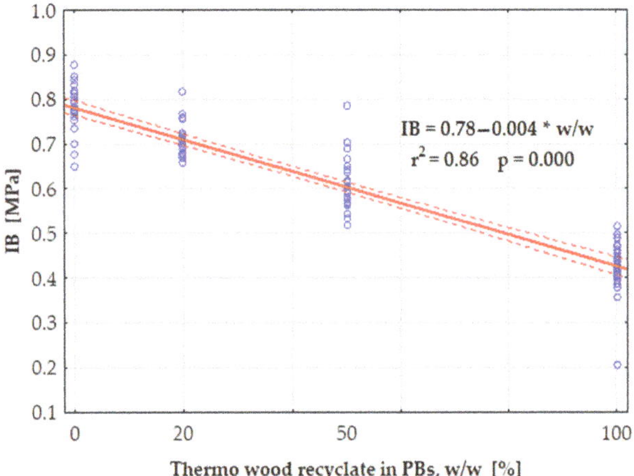

Figure 9. Internal bond (IB) of PBs containing different amount of thermally modified wood recyclates.

The moisture properties of PBs, i.e., the thickness swelling (TS) and water absorption (WA), significantly improved with higher portion of TMW particles in them. For example, TS-2h decreased from 6.01% for PB-C to 1.97% for PB-TMW-100; TS-24h decreased from 23.73% for PB-C to 6.58% for PB-TMW-100; WA-2h decreased from 27.40% for PB-C to 6.38% for PB-TMW-100; and WA-24h decreased from 68.18% for PB-C to 24.55% for PB-TMW-100 (Table 3). The effect of the recycled TMW particles on the moisture properties of PBs was in all cases considered as significant, as values of the coefficient of determination r^2 for the linear correlations "TS or WA = a + b × w/w" were high and ranged between 0.77–0.85, at which p was always 0.000 (Figures 6 and 7). Achieved results related to a better moisture resistance of PBs containing the TMW particles can be explained by their most evident hydrophobic characters compared to particles from fresh spruce wood [12].

Similar results related to moisture properties of solid TMW and PBs based on various TMW types were obtained in more research works, e.g., in Majka et al. [48], Cai et al. [49], Altgen et al. [50]. Ohlmeyer and Lukowsky [24] found out for one layer PBs produced from non-treated and thermally-treated pine wood particles, that the TS-24h was due to presence of TMW reduced from 12.9% to 6.7% and the WA-24h from 59% to 46%. By Borysiuk et al. [4], for PBs based on TMW particles the TS-2h reduced by 20% and the TS-24h by 5%, while the WA-2h reduced by 30% and the WA-24h by 10%. A similar effect of TMW in OSBs on reduction of their TS obtained Paul et al. [29].

Boonstra et al. [25] found out that the WA-24h of one-layer PBs produced from spruce and pine particles pre-heated in two-stages at temperatures below 200 °C was improved only partly. However, some process conditions seem to result in a significant reduction of the WA values, especially at using a higher thermolysis temperature and a longer time of curing process, probably due to a lower sorption capacity of bonded water in the cell walls of wood. By Melo et al. [28], the additional thermal treatments of PBs significantly decreased their WA-24h, the best at using a temperature of 180 °C for six minutes. As for the mechanical properties, the thermal treatments did not significantly influence the PBs strength. Zheng et al. [26] produced PBs from in hot-water pre-treated wood particles, but on the contrary, such PBs had a significantly higher TS and WA, i.e., the TS-24h and WA-24h increased by about 21% and 31%, respectively. Therefore, a hot water pre-treatment of wood particles is not recommended for making of PBs.

Presence of the TMW particles in PBs had a negative effect on their mechanical properties—significantly and very apparently on the modulus of rupture (MOR) and internal bond (IB) with r^2 equal to 0.87 and 0.86 for the linear correlations "MOR or IB = a + b × w/w" (Table 3 and Figures 8a and 9). The TMW particles had also a significantly negative effect on the modulus of elasticity (MOE) of PBs, however, with a lower manifestation when r^2 of the linear correlation "MOE = a + b × w/w" was only 0.19 (Table 3 and Figure 8b).

Compared to the requirements by the standard EN 312 [45], the PBs based on recycled TMWs achieved in all cases the IB 0.35 MPa needed for the type P2.

Generally, results related to lower mechanical properties of PBs containing thermally modified wood are in accordance with work of Ohlmeyer and Lukowsky [24], which determined that the IB equal 0.81 MPa of PBs from non-treated pine wood particles was significantly higher compared to IB equal 0.57 MPa of PBs from thermally treated pine wood particles. Important was also technological process of PBs production, when the PBs from thermally modified wood prepared at pressing factor 8 s/mm had the IB 0.43 MPa, while at 12 mm/s their IB was higher 0.70 MPa. The MOR of PBs prepared from TMW particles was significantly reduced to 14.5 MPa, from 18 MPa of the reference PB.

Borysiuk et al. [4] found out that PBs from TMW particles had a significant decrease of the MOR (about 28%) and the IB (about 22%), while the MOE increased (about 16%). The decrease in the strength properties of MOR and IB was probably related to limitation of surface wettability of the thermal modified particles at the gluing process using UF resins [51]. These authors also state that the increase in the MOE of PBs containing thermally treated wood particles is associated with embrittlement and higher rigidity of these particles, which decrease their deformability in accordance with the knowledge of Hill [2] and also other researchers. In this case, the plasticized lignin stiffens the cell walls in the thermally modified wood particles [6], which resulted in an increase in the stiffness of the entire PB produced from them. A similar effect on the increased MOE of PBs as well as on the shortened lifetime of tools used for machining of PBs could have higher amounts of minerals often presented in recycled woods [52].

3.2. Biological Resistance of PBs

Today, many wood-based composites, including PBs, are produced with no preservative at all. This is partially due to their dominance application in interiors, rather than in outdoor locations with a higher risk of microbial attack. However, for special products from PBs, e.g., used in bathroom or unventilated cellar, it is possible protect them with nano zinc-oxide and other environmentally acceptable biocides [53], using three primary strategies [54,55]: (1) treatment of the wood components prior to PB manufacture; (2) addition of biocides to the adhesives or matrix of the composite; and (3) biocidal posttreatment of PB after its manufacture. The PBs made from recycled TMWs can be considered as more ecofriendly due to the absence of any additional chemicals, at which such thermally modified wood substance has improved also dimensional stability.

TMW displays that the degree of improved decay resistance is positively related to the heating temperature and its duration [56–60]. Unfortunately, the experiences with fungal durability tests for thermally treated composite panel products are still rather limited. Durability tests of the thermally modified strands used for OSB at 200 °C, 220 °C and 240 °C showed improved resistance to the brown-rot fungus *Postia placenta* compared with untreated ones [30]. Similarly, thermally modified wood fibers showed that wood plastic composites produced on their basis had improved resistance against fungi [61,62]. The study of Barnes et al. [63] assessed thermally modified plywood, OSB, laminated strand lumber, and laminated veneer lumber as a posttreatment, using a closed pressurized treatment method, with reduction in mass loss when subjected to a laboratory soil block durability test.

Biodegradation of PBs with the brown-rot fungus *Serpula lacrymans* was more suppressed if the boards contained higher number of particles from the recycled TMWs (Table 4, Figure 10). This result is in accordance with work of [64], by which the resistance of PBs to decaying fungi and termites increased with using of higher number of particles from more durable wood species. The reference PB-C had the greatest mass loss (Δm equal 13.20%), while the PB-TMW-100 containing only particles from TMW offcuts had the smallest mass loss (Δm equal 7.26%)—so the Δm decreased by 45%. The coefficient of determination

r^2 of the linear correlations "Mass loss = a + b × w/w" was 0.67, on the basis of which a less intense decay of PBs containing TMW particles was confirmed with high significance (Figure 10a). Validity of the performed decay test was confirmed by an enough virulence of the fungal strain which at the *Pinus sylvestris* sapwood samples caused mass loss of 26.80%, i.e., more than the level 20% required for this reference wood species by standard EN 113 [65].

Table 4. Biological resistance of the reference/control particleboard (PB-C) and the particleboards (PBs) containing particles from a mixture of recycled thermally modified wood offcuts (PB-TMW): (I.) decay resistance valued on the basis of mass losses (Δm) caused by *S. lacrymans*; (II.) mould resistance against mixture of microscopic fungi valued on the basis of mould growth activity on surfaces of PBs (MGA from 0 to 4).

Biological Resistance of PB	Thermally Modified Wood (TMW) in PB w/w (%)			
	0	20	50	100
Decay attack by *S. lacrymans*				
Δm (%)	13.20 (0.48)	8.42 (0.99)	8.27 (0.72)	7.26 (0.18)
w (%)	87.3 (7.2)	70.1 (5.3)	69.4 (5.2)	68.7 (2.8)
Attack by mixture of moulds MGA (0–4)				
7th day	1.33	1.00	1.00	0.67
14th day	2.33	2.00	2.00	2.00
21st day	3.00	3.00	3.00	3.00
28th day	4.00	4.00	4.00	4.00

Notes: Mean value for mass loss of PB caused by *S. lacrymans* is from 6 samples, and for mould growth activity on the top surface of PB is also from 6 samples. Standard deviations are in the parentheses.

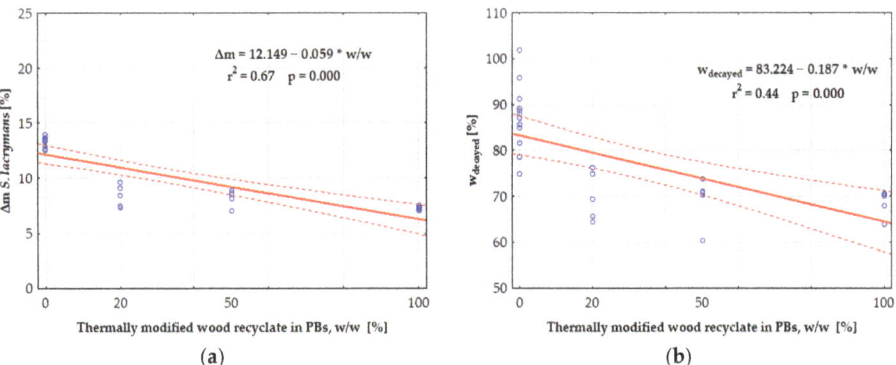

Figure 10. Loss of mass (**a**) and moisture content (**b**) of PBs containing different amount of thermally modified wood recyclates after their attack by the brown-rot fungus *S. lacrymans*.

The moisture content of PBs exposed to *S. lacrymans* was lower for those ones containing TMW particles in comparison to the reference PB (Table 4, Figure 10b), in accordance with their higher hydrophobicity determined by the water absorption test (Table 3, Figure 7). The thermal modification of wood contributed to the increase in resistance of produced PBs to water, when after decay action of *S. lacrymans* the reference PB noted 87.29% moisture content while the PBs containing 100% particles from recycled TMW had the smallest moisture content of 68.69% (decrease about 21.3%).

Similar increase in the decay resistance of PBs and also other wood-based composites containing TMW was obtained by some other researchers. According to the studies [8,57,66–68], decay resistance of TMW to be related to reduction in maximum moisture capacity of the cell wall due to changed and partly degraded wood polymers, mainly hemicelluloses and due to modification of lignin and creation of new linkages in the polysaccharide-lignin matrix of wood. This phenomenon is similar to our previous study when occurred a positive effect of recyclate from faulty PBs on the evidently improved moisture properties and decay resistance to the brown-rot fungus *Coniophora puteana* of newly-prepared PBs [38]. Improved decay resistance of PBs from TMW is in accordance with various types of wood-based composites containing thermally modified components as listed in [30,61,62], as well as wood-based composites exposed to post-heating with reduced mass loss at their following attack by wood decaying fungi [63].

The resistance of PBs to mixture of moulds is summarized in Table 4. The PBs containing 100% TMW recyclates had lower mould rating only on the 7th day of mould test, probably due to reduction the water affinity [8], whereas on the final 28th day after inoculation, 50% and more of their area was covered with the mould hyphae with the sporangium (MGA ranged as 4). The MGA on other PBs containing lower number of particles from recycled TMW and reference PBs had a very similar tendency over time (Table 4).

The differences in the compression strength and FTIR analyses of the used TMWs and spruce wood on properties of prepared PBs can be explained as follows:

- An apparently higher compressive strength parallel with grains of three TMWs (in average 72.1 MPa), compared to less dense spruce wood (42.3 MPa), did not ultimately result in higher MOR of PBs based on TMW particles, but in the opposite effect (Table 3, Figure 8a). It can be attributed to a higher hydrophobicity of TMW particles having a lower polar component of surface free energy [69] with a negative impact on the strength of adhesive joints created between wood elements and polar adhesives [70], including particles in PBs jointed with the polar UF resin.

- FTIR analysis (Table 1) showed that the mixture of three used TMWs, comparing to spruce wood particles, had in accordance with the absorbance peaks an evidently higher portion of lignin determined at 1600 cm^{-1} (1.77:0.75) and unconjugated –C=O groups determined at 1730 cm^{-1} (2.22:1.09), but on the contrary, a lower portion of guaiacyl lignin determined at 1274 cm^{-1} (0.30:1.64) and conjugated –C=C-C=O groups determined at 1653 cm^{-1} (0.06:0.17). Differences in the lignin portion and composition can be attributed: (1) to different wood species (pine/beech/ash TMWs vs. spruce wood) used in the experiment (e.g., generally is known that spruce and other coniferous contain mainly quaiacyl lignin); (2) to creation of new carbonyl and carboxyl groups due to oxidation of wood components in the presence of oxygen at preparation of TMWs; (3) to thermal degradation of hemicelluloses and condensation reactions in TMWs with an indirect increase of lignin and condensed substances similar to natural lignin [71,72].

From the practical point of views, the PBs made with a portion of recycled TMW could be applied for special interior exposures, e.g., kitchens and bathrooms, where sometimes a relatively high humidity of air acts with the aim to protect products made from these boards from moisture and fungal attacks.

4. Conclusions

- The moisture properties of PBs significantly improved with a higher number of hydrophobic TMW particles in their surface and core layers, e.g., after 24 h the thickness swelling (TS) reduced from 23.7% up to 6.58% and the water absorption (WA) from 68.2% up to 24.6%.
- TMW particles had a negative effect on the all mechanical properties of PBs—the most on the modulus of rupture (MOR) in bending with worsening from 14.6 MPa up to 6.5 MPa and the internal bond (IB) with a decrease from 0.78 MPa up to 0.42 MPa, while drop of the modulus of elasticity (MOE) in bending was milder from 2611 MPa maximally to 2189 MPa.
- The biological resistance of PBs to the brown-rot fungus *Serpula lacrymans* significantly increased (maximally about 45%) with a higher portion of more durable TMW particles in their surface and core layers. On the contrary, the presence of TMW particles in PBs had not an evidently positive effect on their mould resistance.

- Generally, adding the recycled TMW particles to produced PBs can be important in terms of increase in their decay and water resistance—required in practice for products exposed to humid environments—but at the same time, worsening of their mechanical properties should be taken into account.

Author Contributions: Conceptualization, J.I., Z.V. and L.R.; methodology, J.I., Z.V. and L.R.; software, J.I. and Z.V.; validation, J.I., Z.V. and L.R.; formal analysis, J.I. and L.R.; investigation, J.I., Z.V. and L.R.; resources, J.I., Z.V. and L.R.; data curation, J.I., Z.V. and L.R.; writing—original draft preparation, J.I., Z.V. and L.R.; writing—review and editing, J.I. and L.R.; visualization, J.I. and Z.V.; supervision, L.R.; project administration, J.I. and L.R.; funding acquisition, L.R. All authors have read and agreed to the published version of the manuscript.

Funding: This work was supported by the Slovak Research and Development Agency under the contract no. APVV-17-0583.

Data Availability Statement: The data presented in this study are available on request from the corresponding author. The data are not publicly available due to privacy restrictions.

Acknowledgments: We would like to thank to the Slovak Research and Development Agency under the contract no. APVV-17-0583 for support of the research published in this article. We also thanks for a providing part of the material for this research to the companies Kronospan, s.r.o., Zvolen, Slovakia and Kronospan CR, spol. s.r.o., Jihlava, Czech Republic.

Conflicts of Interest: The authors declare no conflict of interest.

References

1. Deppe, H.J.; Ernst, K. *Taschenbuch der Spanplattentechnik*, 4th ed.; DRW-Verlag: Leinfelden-Echterdingen, Germany, 2000; 552p.
2. Hill, C.A.S. *Wood Modification—Chemical, Thermal and Other Processes*; John Wiley & Sons: Chichester, UK, 2006; 260p.
3. Jones, D.; Sandberg, D. A review of wood modification globally—Updated findings from COST FP1407. *Interdiscip. Perspect. Built Environ.* **2020**, *1*, 1. [CrossRef]
4. Borysiuk, P.; Chrzanowski, Ł.; Auriga, R.; Boruszewski, P. Thermally modified wood as raw material for particleboard production. *Chip Chipless Woodwork. Process.* **2016**, *10*, 241–245.
5. *ThermoWood Handbook*; International Thermowood Association: Helsinki, Finland, 2003. Available online: www.thermowood.fi (accessed on 4 October 2021).
6. Niemz, P.; Hofmann, T.; Rétfalvi, T. Investigation of chemical changes in the structure of thermally modified wood. *Maderas-Cienc. Tecnol.* **2010**, *12*, 69–78. [CrossRef]
7. Sandberg, D.; Kutnar, A.; Karlsson, O.; Jones, D. *Wood Modification Technologies—Principles, Sustainability, and the Need for Innovation*; Taylor & Francis Group, LLC.: Oxon, UK, 2021; 431p.
8. Sivrikaya, H.; Can, A.; de Troya, T.; Conde, M. Comparative biological resistance of differently thermal modified wood species against decay fungi, *Reticulitermes grassei* and *Hylotrupes bajulus*. *Maderas-Cienc. Tecnol.* **2015**, *17*, 559–570. [CrossRef]
9. Yildiz, S.; Gezer, E.D.; Yildiz, U.C. 2006: Mechanical and chemical behavior of spruce wood modified by heat. *Build. Environ.* **2006**, *41*, 1762–1766. [CrossRef]
10. Sandberg, D.; Kutnar, A.; Mantanis, G. Wood modification technologies—A review. *Iforest-Biogeosci. For.* **2017**, *10*, 895–908. [CrossRef]
11. Ali, M.R.; Abdullah, U.H.; Ashaari, Z.; Hamid, N.H.; Hua, L.S. Hydrothermal modification of wood: A Review. *Polymers* **2021**, *13*, 2612. [CrossRef]
12. Hill, C.; Altgen, M.; Rautkari, L. Thermal modification of wood—A review: Chemical changes and hygroscopicity. *J. Mater. Sci.* **2021**, *56*, 6581–6614. [CrossRef]
13. Hill, C.; Hughes, M.; Gudsell, D. Environmental impact of wood modification. *Coatings* **2021**, *11*, 366. [CrossRef]
14. Spear, M.J.; Curling, S.F.; Dimitriou, A.; Ormondroyd, G.A. Review of functional treatments for modified wood. *Coatings* **2021**, *11*, 327. [CrossRef]
15. Pelaez-Samaniego, M.R.; Yadama, V.; Lowell, E.; Espinoza-Herrera, R. A review of wood thermal pretreatments to improve wood composite properties. *Wood Sci. Technol.* **2013**, *47*, 1285–1319. [CrossRef]
16. Militz, H.; Altgen, M. Processes and properties of thermally modified wood manufactured in Europe. *Deterior. Prot. Sustain. Biomater.* **2014**, *16*, 269–285. [CrossRef]
17. Jones, D.; Sandberg, D.; Goli, G.; Todaro, L. *Wood Modification in Europe: A State-of-the-Art about Processes, Products and Applications*; Firenze University Press: Firenze, Italy, 2019; 123p. [CrossRef]
18. Ninane, M.; Pollet, C.; Hébert, J.; Jourez, B. Physical, mechanical, and decay resistance properties of heat-treated wood by Besson® process of three European hardwood species. *Biotechnol. Agron. Soc. Environ.* **2021**, *25*, 129–139, Available online: https://popups.uliege.be/1780-4507/index.php?id=19050. (accessed on 28 September 2021). [CrossRef]

19. Borysiuk, P.; Mamiński, M.; Grześkiewicz, M.; Parzuchowski, P.; Mazurek, A. Thermally modified wood as raw material for particleboard manufacture. In Proceedings of the 3rd European Conference on Wood Modification, Cardiff, UK, 15–16 October 2007; pp. 227–230.

20. FAOSTAT. Food and Agriculture Organization of the United Nations. *Forestry Production and Trade*. 2021. Available online: http://www.fao.org/faostat/en/#data/FO (accessed on 12 August 2021).

21. Hsu, W.E. Method of Making Dimensionally Stable Composite Board and Composite Board Produced by Such Method. Canadian Patent No. 1215510, 23 December 1986.

22. Hsu, W.E.; Schwald, W.; Shields, J.A. Chemical and physical changes required for producing dimensionally stable wood based composites. Part 1: Steam pretreatment. *Wood Sci. Tehnol.* **1988**, *22*, 281–289. [CrossRef]

23. Sekino, N.; Inoue, M.K.; Irle, M.; Adcock, T.E. The Mechanisms behind the improved dimensional stability of particleboards made from steam-pretreated particles. *Holzforschung* **1999**, *53*, 435–440. [CrossRef]

24. Ohlmeyer, M.; Lukowsky, D. Wood-based panels produced from thermal-treated materials: Properties and perspectives. In *Woodframe Housing Durability and Disaster Issues, Proceedings of the Woodframe Housing Durability and Disaster Issues Conference, Las Vegas, NV, USA, 4–6 October 2004*; Forest Products Society: Madison, WI, USA, 2005; pp. 127–131.

25. Boonstra, M.J.; Pizzi, A.; Zomers, F.; Ohlmeyer, F.; Paul, W. The effects of a two stage heat treatment process on the properties of particleboard. *Holz Als Roh-und Werkst.* **2006**, *64*, 157–164. [CrossRef]

26. Zheng, Y.; Pan, Z.; Zhang, R.; Jenkins, B.M.; Blunk, S. Properties of medium-density particleboard from saline Athel wood. *Ind. Crop. Prod.* **2006**, *23*, 318–326. [CrossRef]

27. Boonstra, M.J. A Two-Stage Thermal Modification of Wood. Ph.D. Dissertation, Ghent University, Ghent, Belgium, Universite Henry Poincare, Nancy, France, 2008; 297p.

28. Melo, R.R.; Muhl, M.; Stangerlin, D.M.; Alfanas, R.F.; Junior, F.R. Properties of particleboards submitted to heat treatments. *Ciênc. Florest.* **2018**, *28*, 776–783. [CrossRef]

29. Paul, W.; Ohlmeyer, M.; Leithoff, H.; Boonstra, M.J.; Pizzi, A. Optimising the properties of OSB by a one-step heat pre-treatment process. *Holz Als Roh-und Werkst.* **2006**, *64*, 227–234. [CrossRef]

30. Paul, W.; Ohlmeyer, M.; Leithoff, H. Thermal modification of OSB-strands by a one-step heat pretreatment—Influence of temperature on weight loss, hygroscopicity and improved fungal resistance. *Holz Als Roh-und Werkst.* **2007**, *65*, 57–63. [CrossRef]

31. Paredes, J.J.; Jara, R.; Shaler, S.M.; van Heiningen, A. Influence of hot water extraction on the physical and mechanical behavior of OSB. *Forest Prod. J.* **2008**, *58*, 56–62.

32. Taylor, A.; Hosseinaei, O.; Wang, S. *Mold Susceptibility of Oriented Strandboard Made with Extractedflakes*; (IRG/WP 08-40402); International Research Group on Wood Protection: Stockholm, Sweden, 2008.

33. Howell, C.; Paredes, J.J.; Jellison, J. Decay resistance properties of hot water extracted oriented strandboard. *Wood Fiber Sci.* **2009**, *41*, 201–208.

34. Paredes, J.J. The Influence of Hot Water Extraction on Physical and Mechanical Properties of OSB. Ph.D. Dissertation, The University of Maine, Orono, ME, USA, 2009.

35. Paredes, J.J.; Shaler, S.M.; Edgar, R.; Cole, B. Selected volatile organic compound emissions and performance of oriented strandboard from extracted southern pine. *Wood Fiber Sci.* **2010**, *42*, 429–438.

36. Mendes, R.F.; Junior, G.B.; Almeida, N.F.; Surdi, P.G.; Barbeiro, I.N. Effect of thermal treatment on properties of OSB panels. *Wood Sci. Technol.* **2013**, *47*, 243–256. [CrossRef]

37. Direske, M.; Bonigut, J.; Wenderdel, C.; Scheiding, W.; Krug, D. Effects of MDI content on properties of thermally treated oriented strand board (OSB). *Eur. J. Wood Prod.* **2018**, *76*, 823–831. [CrossRef]

38. Iždinský, J.; Vidholdová, Z.; Reinprecht, L. Particleboards from recycled wood. *Forests* **2020**, *11*, 1166. [CrossRef]

39. EN 323. *Wood-Based Panels—Determination of Density*; European Committee for Standardization: Brussels, Belgium, 1993.

40. EN 322. *Wood-Based Panels—Determination of Moisture Content*; European Committee for Standardization: Brussels, Belgium, 1993.

41. EN 317. *Particleboards and Fibreboards—Determination of Swelling in Thickness after Immersion in Water*; European Committee for Standardization: Brussels, Belgium, 1993.

42. STN 490164. *Particle Boards—Determination of Water Absorption*; Slovak Office of Standards, Metrology and Testing: Bratislava, Slovak, 1980.

43. EN 310. *Wood-Based Panels—Determination of Modulus of Elasticity in Bending and of Bending Strength*; European Committee for Standardization: Brussels, Belgium, 1993.

44. EN 319. *Particleboards and Fibreboards—Determination of Tensile Strength Perpendicular to the Plane of the Board*; European Committee for Standardization: Brussels, Belgium, 1993.

45. EN 312. *Particleboards—Specifications*; European Committee for Standardization: Brussels, Belgium, 2010.

46. ENV 12038. *Durability of Wood and Wood-Based Products—Wood-Based Panels—Method of Test for Determining the Resistance against Wood-Destroying Basidiomycetes*; European Committee for Standardization: Brussels, Belgium, 2002.

47. EN 15457. *Paints and Varnishes—Laboratory Method for Testing the Efficacy of Film Preservatives in a Coating against Fungi*; European Committee for Standardization: Brussels, Belgium, 2007.

48. Majka, J.; Roszyk, E. Swelling restraint of thermally modified ash wood perpendicular to the grain. *Eur. J. Wood Wood Prod.* **2018**, *76*, 1129–1136. [CrossRef]

49. Cai, C.; Javed, M.A.; Komulainen, S.; Telkki, V.; Haapala, A.; Heräjärvi, H. Effect of natural weathering on water absorption and pore size distribution in thermally modified wood determined by nuclear magnetic resonance. *Cellulose* **2020**, *27*, 4235–4247. [CrossRef]

50. Altgen, D.; Altgen, M.; Kyyrö, S.; Rautkari, L.; Mai, C. Time-dependent wettability changes on plasma-treated surfaces of unmodified and thermally modified European beech wood. *Eur. J. Wood Wood Prod.* **2020**, *78*, 417–420. [CrossRef]

51. Boruszewski, P.; Borysiuk, P.; Mamiński, M.Ł.; Grześkiewicz, M. Gluability of thermally modified beech (*Fagus silvatica* L.) and birch (*Betula pubescens* Ehrh.) wood. *Wood Mater. Sci. Eng.* **2011**, *6*, 185–189. [CrossRef]

52. Patsch, R.; Frömel-Frybort, S.; Stanzl-Tschegg, S.E. The influence of the recycled wood proportion in particle boards to the tool life of milling tools. *Wood Mater. Sci. Eng.* **2020**. [CrossRef]

53. Reinprecht, L.; Iždinský, J.; Vidholdová, Z. Biological resistance and application properties of particleboards containing nano-zinc oxide. *Adv. Mater. Sci. Eng.* **2018**, *2018*, 2680121. [CrossRef]

54. Spear, M.J. Preservation, protection and modification of wood composites. *Wood Compos.* **2015**, 253–310.

55. Reinprecht, L. *Wood Deterioration, Protection and Maintenance*; John Wiley & Sons, Ltd.: Chichester, UK, 2016; 357p.

56. Kamdem, D.P.; Pizzi, A.; Jermannaud, A. Durability of heat-treated wood. *Holz Als Roh-und Werkst.* **2002**, *60*, 1–6. [CrossRef]

57. Rapp, A.O.; Brischke, C.; Welzbacher, C.R.; Jazayeri, L. Increased resistance of thermally modified Norway spruce timber (TMT) against brown rot decay by *Oligoporus placenta*—Study on the mode of protective action. *Wood Res.* **2008**, *53*, 13–25.

58. Reinprecht, L.; Vidholdová, Z. Mould resistance, water resistance and mechanical properties of OHT-thermowoods. In Proceedings of the Sustainability through New Technologies for Enhanced Wood Durability: Socio-Economic Perspectives of Treated Wood for the Common European Market, Cost Action E37 Final Conference, Bordeaux, France, 29–30 September 2008; Acker, Peek; Ghent University: Ghent, Belgium, 2008; pp. 159–165.

59. Allegretti, O.; Brunetti, M.; Cuccui, I.; Ferrari, S.; Nocetti, M.; Terziev, N. Thermo-vacuum modification of spruce (*Picea abies* Karst.) and fir (*Abies alba* Mill.) wood. *BioResources* **2012**, *7*, 3656–3669.

60. Chaouch, M.; Dumarçay, S.; Pétrissans, A.; Pétrissans, M.; Gérardin, P. Effect of heat treatment intensity on some conferred properties of different European softwood and hardwood species. *Wood Sci. Technol.* **2013**, *47*, 663–673. [CrossRef]

61. Kuka, E.; Cīrule, D.; Kajaks, J.; Janberga, A.; Andersone, I.; Andersons, B. Fungal degradation of wood plastic composites made with thermally modified wood residues. *Key Eng. Mater.* **2017**, *721*, 8–12. [CrossRef]

62. Westin, M.; Larsson-Brelid, P.; Segerholm, B.K.; Oever, M. *Wood Plastic Composites from Modified Wood. Part 3. Durability of WPCs with Bioderived Matrix*; (IRG/WP 08-40423); International Research Group on Wood Protection: Stockholm, Sweden, 2008.

63. Barnes, H.M.; Aro, M.D.; Rowlen, A. Decay of thermally modified engineered wood products. *For. Prod. J.* **2018**, *68*, 99–104.

64. Behr, E.A.; Wittrup, B.A. Decay and termite resistance of two species of particleboards. *Holzforschung* **1969**, *23*, 166–170. [CrossRef]

65. EN 113. *Wood Preservatives. Test Method for Determining the Protective Effectiveness Against Wood Destroying Basidiomycetes. Determination of the Toxic Values*; European Committee for Standardization: Brussels, Belgium, 1996.

66. Thybring, E.E. The decay resistance of modified wood influenced by moisture exclusion and swelling reduction. *Int. Biodeterior Biodegrad.* **2013**, *82*, 87–95. [CrossRef]

67. Weiland, J.J.; Guyonnet, R. Study of chemical modifications and fungi degradation of thermally modified wood using DRIFT spectroscopy. *Holz Als Roh-und Werkst.* **2003**, *61*, 216–220. [CrossRef]

68. Ringman, R.; Beck, G.; Pilgård, A. The importance of moisture for brown rot degradation of modified wood: A critical discussion. *Forests* **2019**, *10*, 522. [CrossRef]

69. Kúdela, J.; Lagaňa, R.; Andor, T.; Csiha, C. Variations in beech wood surface performance associated with prolonged heat treatment. *Acta Fac. Xylol. Zvolen* **2020**, *62*, 5–17. [CrossRef]

70. Vidholdová, Z.; Ciglian, D.; Reinprecht, L. Bonding of the thermally modified Norway spruce wood with the PUR and PVAc adhesives. *Acta Fac. Xylol. Zvolen* **2021**, *63*, 63–73. [CrossRef]

71. Sikora, A.; Kačík, F.; Gaff, M.; Vondrová, V.; Bubeníková, T.; Kubovský, I. Impact of thermal modification on color and chemical changes of spruce and oak wood. *J. Wood Sci.* **2018**, *64*, 406–416. [CrossRef]

72. Výbohová, E.; Kučerová, V.; Andor, T.; Balážová, Ž.; Veľková, V. The effect of heat treatment on the chemical composition of ash wood. *BioResources* **2018**, *13*, 8394–8408. [CrossRef]

Particleboards from Recycled Pallets

Ján Iždinský, Ladislav Reinprecht and Zuzana Vidholdová

Department of Wood Technology, Faculty of Wood Sciences and Technology, Technical University in Zvolen, T. G. Masaryka 24, 96001 Zvolen, Slovakia; reinprecht@tuzvo.sk (L.R.); zuzana.vidholdova@tuzvo.sk (Z.V.)
* Correspondence: jan.izdinsky@tuzvo.sk

Abstract: Worldwide production of wooden pallets continually increases, and therefore in future higher number of damaged pallets need to be recycled. One way to conveniently recycle pallets is their use for the production of particleboards (PBs). The 3-layer PBs, bonded with urea-formaldehyde (UF) resin, were prepared in laboratory conditions using particles from fresh spruce logs (FSL) and recycled spruce pallets (RSP) in mutual weight ratios of 100:0, 80:20, 50:50 and 0:100. Particles from RSP did not affect the moisture properties of PBs, i.e., the thickness swelling (TS) and water absorption (WA). The mechanical properties of PBs based on particles from RSP significantly worsened: the modulus of rupture (MOR) in bending from 14.6 MPa up to 10 MPa, the modulus of elasticity (MOE) in bending from 2616 MPa up to 2012 MPa, and the internal bond (IB) from 0.79 MPa up to 0.61 MPa. Particles from RSP had only a slight negative effect on the decay resistance of PBs to the brown-rot fungus *Serpula lacrymans*, while their presence in surfaces of PBs did not affect the growth activity of moulds at all.

Keywords: particleboards; pallets; recycled wood; physical and mechanical properties; decay and mould resistance; FTIR

Citation: Iždinský, J.; Reinprecht, L.; Vidholdová, Z. Particleboards from Recycled Pallets. *Forests* **2021**, *12*, 1597. https://doi.org/10.3390/f12111597

Academic Editor: Antonios Papadopoulos

Received: 16 October 2021
Accepted: 16 November 2021
Published: 19 November 2021

Publisher's Note: MDPI stays neutral with regard to jurisdictional claims in published maps and institutional affiliations.

1. Introduction

Today, an increasing emphasis is placed not only on the recycling of municipal waste but also on other assortments, such as wood, which can be reused for the production of wooden agglomerated materials, especially particleboards (PBs). The measures introduced by the EU seek to close the life cycle of products and materials by preserving, as far as possible, their value to the economy, minimizing waste production and maximizing recycling and reusing. Thus, the benefits are mainly in the environment and the economy [1,2].

The idea of wood waste recycling has only started to be taken more seriously in recent decades, as seen in the work of Ihnát et al. [2]; in the past, researchers have dealt with the issue of wood recycling predominantly at the theoretical and laboratory levels. However, application in industry has been progressively and over time, considering more economic [3] and environmental benefits [4–6]. Today, old waste wooden products—mainly pallets, drums and furniture from solid wood and wooden composites (PBs, oriented strand boards (OSBs), plywood, etc.)—represent a significant resource for manufacturing new PBs [2,7].

Accordingly, in the German Decree for material recycling and thermal recycling, waste wood is categorized into four classes (A I to A IV), found in the attachments of the German Waste Wood Ordinance [8]. Pallets are divided into four categories: (I) Pallets made of solid wood, (II) Pallets of composite wood materials, (III) Other pallets with composite materials, (IV) Wood pallets with preservatives and other wood with high pollutant content intended for energy use only [8].

Pallets are used for storing, protecting, and transporting freight. They are the most common base for handling and moving the unit load, carried by materials handling units, such as forklifts [9]. The pallet market continually increases due to the rising standard of goods transportation, the adoption of modern material handling units in different industries, and market demand for palletized goods. More than 600 million Euro-pallets are in circulation in the global logistics industry. In 2020, 123.5 million wooden the European Pallet Association e.V. (EPAL) pallets and other carriers were produced, which is 0.5 million more compared to 2019 [10].

Approximately 450–500 million new pallets are manufactured annually and join approximately 2 billion pallets that are in circulation in the U.S. [11]. In the European Union, some 280 million pallets are in circulation every year [12].

The two most common pallet types in the European market are the Euro-pallet and the industrial pallet. Both pallet types are standardized and used in the European transport market. The Euro-pallet has a length of 1.2 m and a width of 0.8 m with a weight of 25 kg when made of wood. The industrial pallet length amounts to 1.2 m and its width is 1.0 m with a weight of 35 kg. The EPAL indicates the load capacity for Euro-pallets of 1500 kg and for industrial pallets of 1250 kg [13].

Pallets are commonly made of wood, which is one of the raw materials often used for containers and is the world's most important renewable material and regenerative fuel [14,15]. Due to the fact that packaging material has already fulfilled its function at the beginning of the use phase of the respective product and is then turned into waste, the environmental relevance of containers materials has become very important [16]. The production of PBs only from recycled pallets of a defined type has not been the subject of much research, because in practice, the pallets usually represent a mix of various types and are made from different wood species [17].

The aim of this work was to study the effect of recycled wood particles—recyclates from recycled spruce pallets (RSP) added into PBs together with various amounts of spruce particles prepared from fresh spruce logs (FSL)—on the selected physical, mechanical and biological properties of PBs.

2. Materials and Methods

2.1. Materials

2.1.1. Wood Particles

In the experiment, two types of wood particles prepared from the fresh spruce logs (FSL = C/control/) and the recycled spruce pallets (RSP) were used. Both wood materials were obtained and processed in the company Kronospan s.r.o., Zvolen, Slovakia (Figure 1a). Selected mechanical properties of the FSL were partly higher than those of the RSP: A) the compression strength parallel to the grain ($\sigma_{Compression\parallel}$) by STN 49 0110 [18] for 15 sample-replicates was 42.3 MPa and 38.2 MPa, respectively; B) the dynamic modulus of elasticity (MOE_d) measured with the ultrasonic timer Fakopp for 12 slab-replicates at their moisture content of app. 30% was 17.73 GPa and 14.51 GPa (Figure 1b), respectively.

From the FSL, chips were prepared on drum chipper TH/N 1000/1250/15 (Rudnick & Enners Maschinen- und Anlagenbau GmbH, Alpenrod, Germany). From the RSP, chips were prepared using the mobile chipper HEM (JENZ GmbH, Petershagen, Germany) combined with the magnetic separation of ferrous metals (iron, steel, etc.), i.e., nails and other connectors on pallets. Following this, wood particles were prepared from wood chips using the Knife Ring Flakers G24 (GOOS Engineering spol. s.r.o., Brno, Czech Republic).

Finally, spruce wood particles were individually milled on finer particles for core and surface layers in the laboratories of the Technical University in Zvolen, using the grinding mill SU1 (TMS, Pardubice, Czech Republic). The dimensions of particles selected for the core layer of PBs were from 0.25 to 4.0 mm, and for the surface layers from 0.125 to 1.0 mm (Figure 2). Particles for the core layer were dried to a moisture content of 2%, and for the surface layers to a moisture content of 4%.

 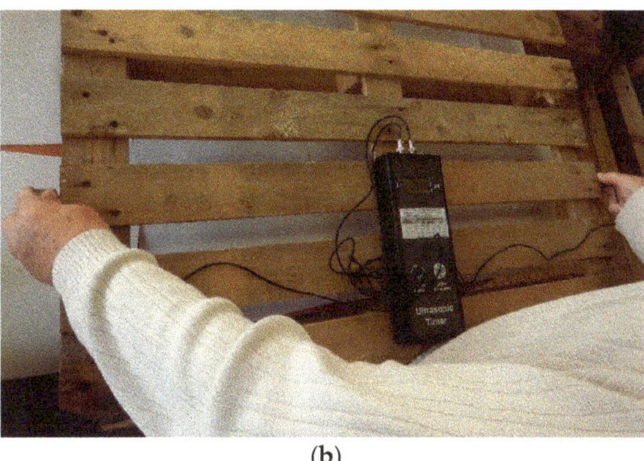

(a) (b)

Figure 1. Views on the storage of discarded pallets in Kronospan s.r.o., Zvolen (**a**), and on the determination of the modulus of elasticity (MOE$_d$ = density × /speed of ultrasonic waves in the longitudinal direction/^{2}) of slab in pallet by the ultrasonic timer Fakopp (**b**).

 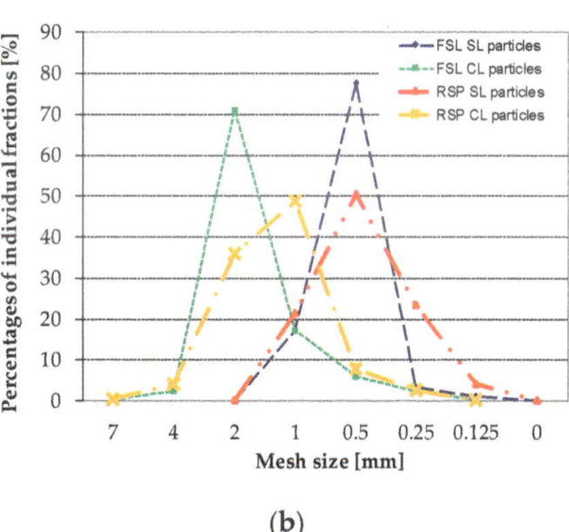

(a) (b)

Figure 2. Particles for PBs: (**a**) Prepared from the fresh spruce logs (FSL)—type I., and from the recycled spruce pallets (RSP)—type II., at which both particle types were used in the surface layers (SL) and the core layer (CL) of PBs; (**b**) Size characteristics of particles from FSL and RSP, with percentage of individual fractions for surface layers (SL) and core layer (CL).

Particles with a dimension of ≤0.25 mm were subjected to FTIR spectra analysis using a Nicolet iS10 spectrometer equipped with Smart iTR ATR accessory using diamond crystal. For both particle types, 4 spectral measurements were performed in the range from 4000 cm^{-1} to 650 cm^{-1} with a resolution of 4 cm^{-1}. Measured spectra were baseline corrected and analyzed in absorbance mode by OMNIC 8.0 software (Table 1).

2.1.2. Resin and Additives

Two urea-formaldehyde (UF) resin types, produced in company Kronospan, were used for mixing with wood particles: (a) KRONORES CB 4005 D for the surface layers of PBs added to particles in an amount of 11 wt.%, and (b) KRONORES CB 1637 D for the core layer of PBs added to particles in an amount of 7 wt.%. The basic characteristics of these two UF resins were as follows: molar ratio of urea to formaldehyde constantly 1 to 1.2; solid content 65.9% and 67.4% by EN 827 [19]; viscosity 76 s and 86 s using Ford cup 4 mm/20 °C by EN ISO 2431 [20]; pH value 9.08 and 8.62 by EN 1245 [21]; gel time 81 s and 36 s by chloride test, respectively. The NH$_4$NO$_3$, used as 57 wt.% water solution, was applied as a hardener for both UF resins, added to their dry mass in amount of 2% and

4%, respectively. Paraffin, used as 35 wt.% water emulsion, was applied on the surface and core particles in amounts of 0.6% and 0.7%, respectively. A higher amount of hardener in the UF resin for the core layer of PB (4%) was needed due to technological conditions of the PBs preparation, i.e., because the temperature increase (needed for curing of UF resin) starts later in core layer as in the surface layers of PBs.

Table 1. Intensity of FTIR spectra (normalized at 898 cm^{-1}) for particles from the fresh spruce logs (FSL) and the recycled spruce pallets (RSP).

FTIR (cm^{-1})	FSL	RSP
1274	1.64	1.24
1334	0.15	0.13
1372	1.11	1.15
1430	1.06	1.09
1510	2.42	2.28
1600	0.75	0.51
1653	0.17	0.18
1730	1.09	1.28
2900	1.60	1.63
$TCI = 1372/2900$	0.69	0.71
$LOI = 1430/898$	1.06	1.09

2.2. Particleboard Preparation

The 3-layer PBs with the dimensions of 400 mm $\times$ 300 mm $\times$ 16 mm were prepared in laboratories of the Technical University in Zvolen. The ratio of surface/core wood particles for all PB-types was constant 35:65. The moisture content of particles mixed with UF resins was 8.5–10.2% in the surface layers (SL), and 5.9–6.9% in the core layer. The PBs were prepared by the same technology as used Iždinský et al. [17,22]—i.e., firstly cold pre-pressing of particle mats at 1 MPa, followed by hot pressing in pressure (CBJ 100–11 laboratory press, TOS, Rakovník, Czech Republic) at a maximum temperature of the pressing plates in the press 240 °C, a maximum pressing pressure of 5.75 MPa, and a pressing factor of 8 s/mm (total pressing time of 128 s). From each PB type, 6 boards were produced, i.e., 24 PBs in total (Table 2).

Table 2. Individual types of manufactured particleboards (PBs).

Variant of PB	Recycled Spruce Pallets (RSP) in PB w (Recycled Wood)/w (Total Wood) (%)	Number of Produced Boards	Board Type
PB-C: 100% particles from fresh spruce logs (FSL)	0	6	C
PB-RSP: 20%, 50% or 100% particles from recycled spruce pallets (RSP), combined with FSL particles	20	6	RSP-20
	50	6	RSP-50
	100	6	RSP-100

2.3. Properties of PBs-Physical, Mechanical, and Biological

Selected physical, mechanical and biological properties of PBs were determined on samples cut out from PBs (Figure 3), in accordance with European (EN) and Slovak (STN) standards.

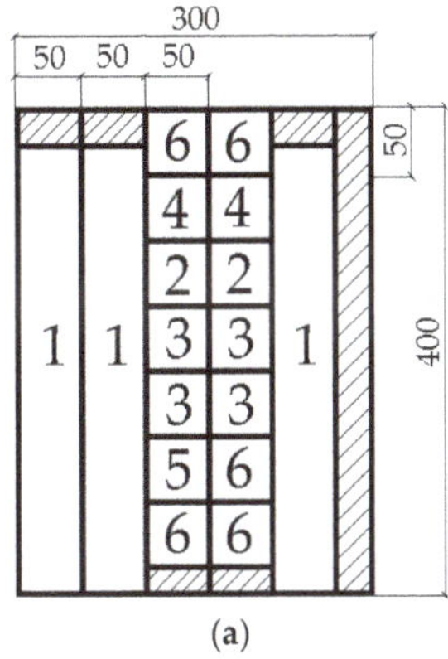 

(a)   (b)

Figure 3. Scheme of samples preparation from the PB (**a**): 1—MOR and MOE by 3-point bending test [23], 2—TS and WA after 2 and 24 h [24,25], 3—IB [26] and density [27], 4—decay resistance to the fungus *Serpula lacrymans* [28], 5—mould resistance [29], 6—spare samples; (**b**) Display of PB-samples 50 mm × 50 mm × 16 mm with different amount of particles from recycled spruce pallets (RSP) *w/w* (%): (A) C = 0%, (B) 20RSP = 20%, (C) 50RSP = 50%, and (D) 100RSP = 100%.

Physical properties: the density by EN 323 [27], the moisture content (w) by EN 322 [30], and the thickness swelling (TS) and water absorption (WA) after 2 and 24 h by EN 317 [24] and STN 490164 [25] were determined.

Mechanical properties using the universal machine TiraTest 2200 (VEB TIW, Rauenstein, Germany): the modulus of rupture (MOR) and modulus of elasticity (MOE) in bending by EN 310 [23]; and the internal bond (IB) by EN 319 [26] were determined as well.

Biological properties: the decay resistance to the brown-rot fungus *Serpula lacrymans* (Schumacher ex Fries) S.F.Gray/*S. lacrymans* (Wulfen) J. Schröt., by IndexFungorum/, strain BAM 87 (Bundesanstalt für Materialforshung und -prüfung, Berlin) performed according to a partly modified ENV 12038 [28] (changes to [28] listed by Iždinský et al. [22]) —determining after 16 weeks the mass loss (Δm) and moisture content ($w_{decayed}$) of PBs; and the mould growth activity (MGA) of the mixture of microscopic fungi (*Aspergillus versicolor* BAM 8, *Aspergillus niger* BAM 122, *Penicillium purpurogenum* BAM 24, *Stachybotrys chartarum* BAM 32 and *Rhodotorula mucilaginosa* BAM 571) on the top surface of PBs by a partly modified EN 15457 [29] (changes to [29] listed by Iždinský et al. [22])—using these criteria: 0 = without mould growth; 1 = mould up to 10%; 2 = mould up to 30%; 3 = mould up to 50%; and 4 = mould more than 50%.

2.4. Statistical Analyses

By the statistical software STATISTICA 12 (StatSoft, Inc., Tulsa, OK, USA) were analyzed the gathered data (the arithmetic mean and standard deviation) and stated the simple linear correlations with the coefficient of determination (r^2) and the probability value (*p*-value) that measures the likelihood that a test-statistic value could occur by chance, given that the null hypothesis is true.

3. Results and Discussion

3.1. Physical and Mechanical Properties of PBs

For the final moisture and strength properties of composites, including PBs, two basic factors are important: (a) structure and properties of the input materials; (b) technological processes of the composite preparation. The moisture and strength properties of PBs are significantly influenced by the type, amount and strength of chemical bonds (etheric and other types of covalent bonds created between the wood components and the resin) and physico–chemical interactions (hydrogen bonds and van der Walls interactions created in the mixture of wood substrate and the resin) by which are in them mutually connected the individual wood particles [31,32].

The basic physical and mechanical properties of PBs are presented in Table 3. The average density of all four PB-types ranged in a very narrow interval from 651 kg·m^{-3} to 657 kg·m^{-3} (Table 3, Figure 4). The densities of the laboratory prepared PBs were comparable with the density of commercial PBs, at which they were not affected by different sources of spruce wood from which the particles for the core and surface layers were prepared.

Table 3. Physical and mechanical properties of the reference/control PB (PB-C) and the PBs containing particles from recycled spruce pallets (PB-RSP).

Property of PB		Recycled Spruce Pallets (RSP) in PB *w/w* (%)			
		0	20	50	100
Density	[kg·m^{-3}]	656 (15.7)	651 (18.1)	657 (21.8)	653 (26.0)
Thickness swelling (TS) after 2 h	[%]	6.00 (0.53)	5.27 (0.27)	11.02 (1.63)	6.47 (0.84)
Thickness swelling (TS) after 24 h	[%]	23.81 (1.38)	18.67 (1.58)	27.87 (1.63)	23.67 (1.25)
Water absorption (WA) after 2 h	[%]	27.43 (2.04)	18.49 (0.92)	41.11 (3.62)	20.62 (0.84)
Water absorption (WA) after 24 h	[%]	68.31 (2.32)	50.95 (2.30)	76.80 (2.24)	56.77 (2.82)
Internal bond (IB)	[MPa]	0.79 (0.06)	0.70 (0.05)	0.68 (0.04)	0.61 (0.03)
Modulus of rupture (MOR)	[MPa]	14.6 (1.56)	12.1 (0.97)	12.4 (1.02)	10.0 (1.27)
Modulus of elasticity (MOE)	[MPa]	2616 (286)	2471 (390)	2276 (248)	2012 (193)

Notes: Mean values of density are from 42 samples, of TS from 12 samples, of WA from 12 samples, of IB from 24 samples, of MOR and MOE from 18 samples. Standard deviations are in the parentheses. The commercial PB type P2 (i.e., by EN 312 [33] boards in interior, including furniture, exposed to dry conditions) has these minimal limits of searched mechanical properties: IB = 0.35 MPa, MOR = 11 MPa, MOE = 1600 MPa.

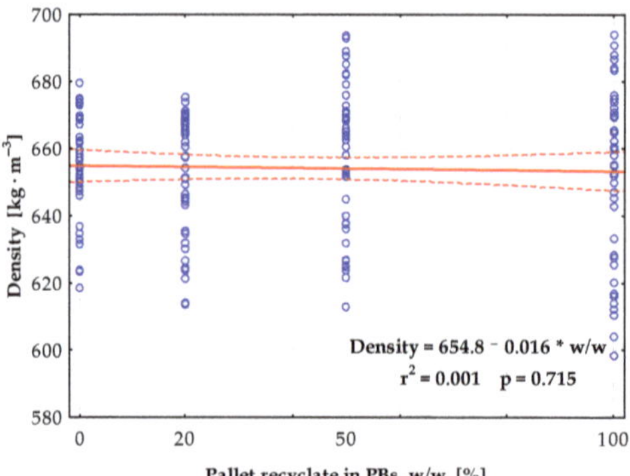

Figure 4. Density of PBs containing different amount of particles from recycled pallets.

The particles from RSP had only a negligible impact on the moisture properties of PBs, i.e., the TS and WA (Table 3, Figures 5 and 6). It is evident from the linear correlations "TS or WA = a + b × *w/w*" between the percentage weight content of RSP particles in PBs (*w/w*) and the TS and WA values of PBs, which were characterized by very small coefficients of determination r^2 from 0.004 to 0.07, and with high p-values from 0.077 to 0.664.

Theoretically, no significant effect of the type of spruce particles in PBs on their TS and WA values can be explained by a similar chemical composition (e.g., amount of amorphous and crystal cellulose; type and amount of hemicelluloses, lignin and extracts) and chemical-physical characteristics (e.g., polymerization degree of cellulose; amount, space accessibility and reactivity of hydroxyl groups) of wood substance presented in the FSL and RSP. The FTIR analyses indicated that the basic chemical composition of wood particles obtained from the FSL and RSP was almost the same (Table 1). Absorbance (A) peaks of both types of spruce particles were usually very similar, i.e., the 1334 cm^{-1}

(specific for syringyl lignin [34]), the 1372 cm^{-1} (belonging to CH deformations in cellulose and hemicelluloses), the 1430 cm^{-1} (assigned to aromatic skeletal vibrations, and to C-H plane vibration in plane cellulose [35], the 1510 cm^{-1} (C=C stretching of the aromatic skeletal vibrations in lignin [36]), and the 1653 cm^{-1} (belonging to conjugated carbonyls –C=C-C=O [37]). Subsequently, by comparing the absorbance ratios in the FSP and RSP particles, the crystallinity indexes of cellulose [38], i.e., the "TCI" (total crystallinity index = A1372/A2900 = 0.69–0.71) and also the "LOI" (lateral order index, or empirical crystallinity index = A1430/A898 = 1.06–1.09), were almost identical for both types of spruce particles. On the contrary, the particles from RSP had a lower absorbance at 1274 cm^{-1}, about 24.4% (specific for quaiacyl lignin), and at 1600 cm^{-1}, about 32% (belonging to aromatic skeletal vibration in lignin, and –C=O stretching [34]), as well as a partly higher absorbance at 1730 cm^{-1}, about 17.4% (stretch of unconjugated –C=O groups of aldehydes, ketones, carboxylic acids and esters in lignin and hemicelluloses [36,39]). Differences in these three A peaks (1274, 1600, and 1730 cm^{-1}) can be attributed to more factors in recycled pallets, e.g., to a local white-rot and a long-term atmospheric oxidation.

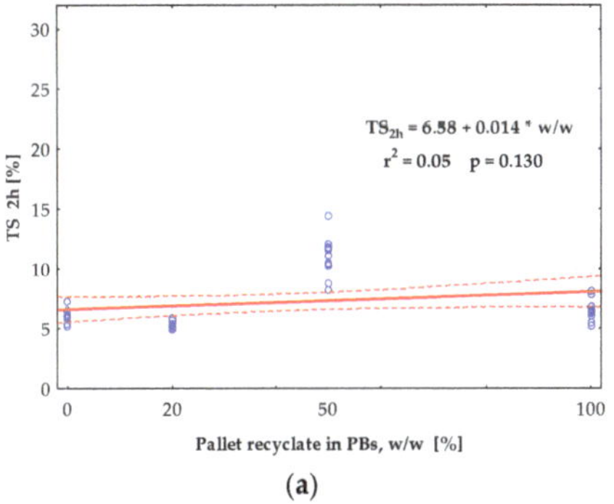
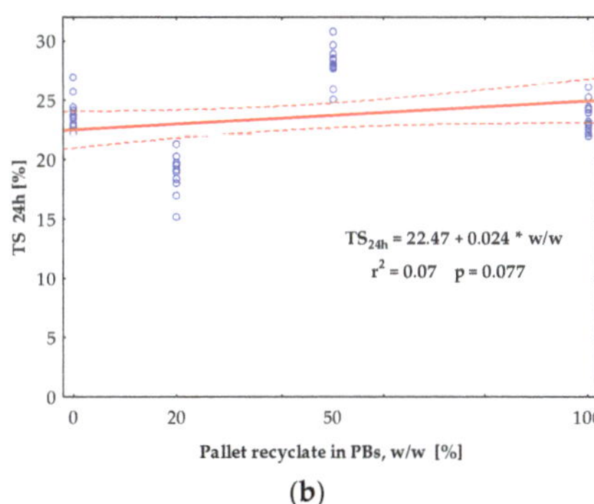

Figure 5. Thickness swelling (TS) after 2 h (**a**) and 24 h (**b**) of PBs containing different amount of particles from recycled spruce pallets (RSP).

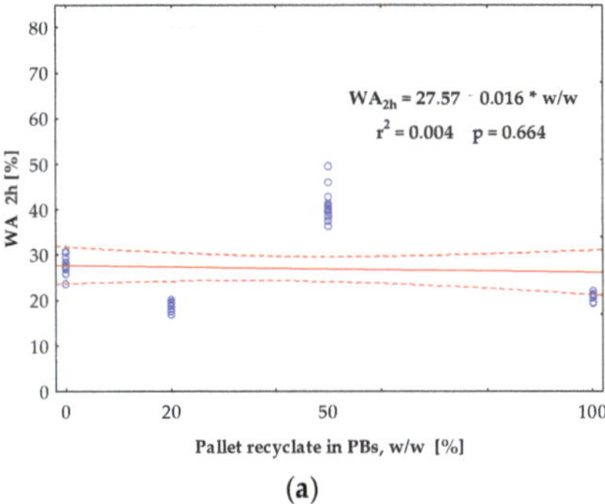
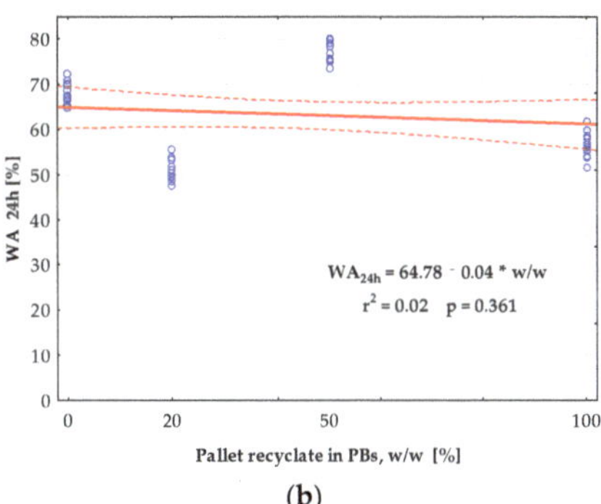

Figure 6. Water absorption (WA) after 2 h (**a**) and 24 h (**b**) of PBs containing different amount of particles from recycled spruce pallets (RSP).

Achieved results point to a similar chemical structure in both types of spruce particles that cannot be generalized, because in practice the wood structure (chemical, anatomical, geometry) of fresh logs and similarly of recycled pallets can be deteriorated differently and into various degrees by biological pests (decaying fungi, insects, etc.) and abiotic effects (fire, UV irradiation, etc.). It also can be observed that the degradation processes in wood sometimes has an undefinable effect on its moisture properties.

No significant effect of particles from the recycled spruce wood (RSP) on the moisture properties (TS and WA) of tested PBs could, in theory, be explained by a similar chemical composition (cellulose, hemicelluloses, lignin, extracts) and physical–chemical properties (e.g., amount, accessibility and reactivity of hydroxyl groups) of the fresh and older spruce woods used in the experiment. However, these results and evaluations cannot be general, because in practice the molecular and anatomical structure of spruce wood in fresh logs and similarly in recycled pallets can be deteriorated differently and into various ranges—e.g., in forest, transport, in stock, during exposure.

The mechanical properties of PBs were negatively affected by the presence of particles from the recycled spruce pallets—i.e., comparing to the reference/control PB which contained particles only from the fresh spruce logs—the MOR decreased by about 31.5% from 14.6 MPa to 10 MPa, the MOE about 23.1% from 2616 MPa to 2012 MPa, and the IB by about 22.8% from 0.79 MPa to 0.61 MPa (Table 3, Figures 7 and 8). In the linear correlations "MOR, MOE, or IB = a + b × w/w", the coefficients of determination r^2 ranged from 0.43 to 0.63 (Figures 7 and 8) and p-values were 0.000, confirming a significant negative effect of particles from the old spruce pallets on the searched mechanical properties of PBs. This result could be attributed to presence of some portion of deteriorated or polluted wood in PBs based on particles from the RSP, as it is well known that wood attacked by decaying fungi or aggressive chemicals has lower strength, elasticity and hardness [40,41]. Searched mechanical properties of samples or slabs from RSP were lower by 10% ($\sigma_{Compression\parallel}$) and by 18% ($MOE_d$) than those from FSL (see point Section 2.1.1).

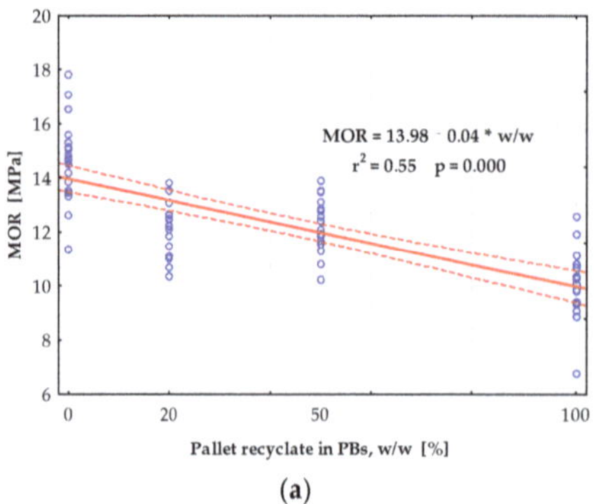

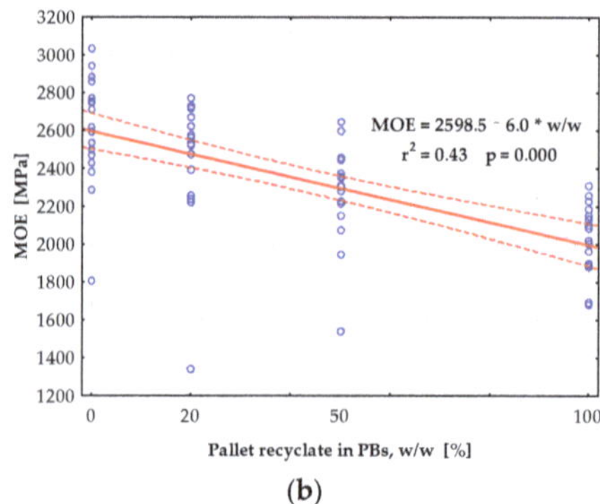

Figure 7. Modulus of rupture (MOR) (**a**) and modulus of elasticity (MOE) (**b**) of PBs containing particles from recycled spruce pallets (RSP).

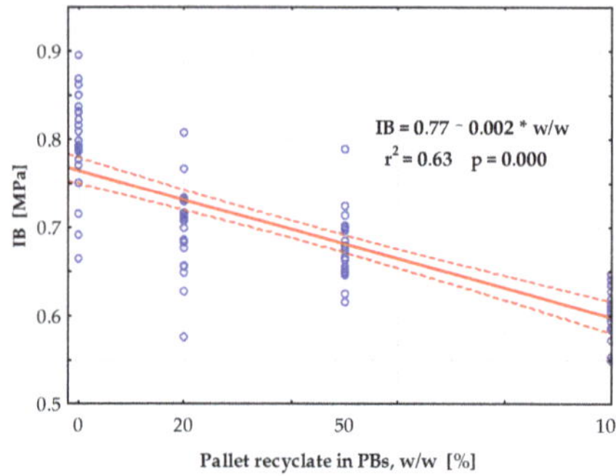

Figure 8. Internal bond (IB) of PBs containing different amount of particles from recycled spruce pallets (RSP).

Generally, results related to lower mechanical properties of PBs containing particles from recycled spruce pallets are in accordance with [42–46]. Compared to the requirements by the standard EN 312 [33], the PBs based on RSP achieved in all cases the IB 0.35 MPa needed for the particleboard type P2.

3.2. Biological Resistance of PBs

The decay attack of PBs with the brown-rot fungus *Serpula lacrymans* was only slight, with a maximum of about 15.4%—inhibited at a higher portion of particles from RSP (Table 4, Figure 9). Due to the 16-week action of *S. lacrymans*, the average mass loss (Δm) of the reference PB-C was equal to 13.20%, while the PB-RSP-100 (containing only particles from RSP) was slightly higher equal to 15.23%. The coefficient of determination r^2 of the linear correlation "Mass loss = a + b × w/w" was 0.325, and the *p*-value was 0.002. These parameters confirmed only with moderate significance a higher intense rot process in those PBs which contained particles from recycled pallets (Figure 9a). After decay tests, the highest moisture content ($w_{decayed}$) had the PB-RSP-100 with the maximum portion of particles from RSP. It indirectly pointed to their more pronounced accessibility for rot (Figure 9b). The performed decay tests seemed to be valid because the strain of *S. lacrymans* caused 26.80% mass loss in the solid *Pinus sylvestris* sapwood, i.e., more than 20% required by the standard EN 113 [47].

Table 4. Biological resistance of the reference/control PB (PB-C) and the PBs containing particles from recycled spruce pallets (PB-RSP): (I) decay resistance valued on the basis of mass loss (Δm) caused by *S. lacrymans*; (II) mould resistance to against a mixture of microscopic fungi valued on the basis of mould growth activity (MGA) on the top surface of PBs (MGA from 0 to 4).

Biological Resistance of PB	Recycled Spruce Pallets (RSP) in PB w/w (%)			
	0	20	50	100
Decay attack by S. lacrymans				
Δm [%]	13.20 (0.48)	12.34 (0.56)	12.77 (0.66)	15.23 (1.61)
w [%]	87.29 (7.23)	89.94 (2.18)	93.41 (5.28)	114.11 (5.85)
Attack by mixture of moulds (MGA [0–4])				
7th day	1.33	1	1	1
14th day	2.33	2	2	2
21st day	2.67	2.33	2.33	3
28th day	4	3.67	3.67	4

Notes: Mean values of mass loss (Δm) in PB caused by *S. lacrymans*, as well as of mould growth activity (MGA) on the top surface of PB are from 6 samples. Standard deviations are in the parentheses.

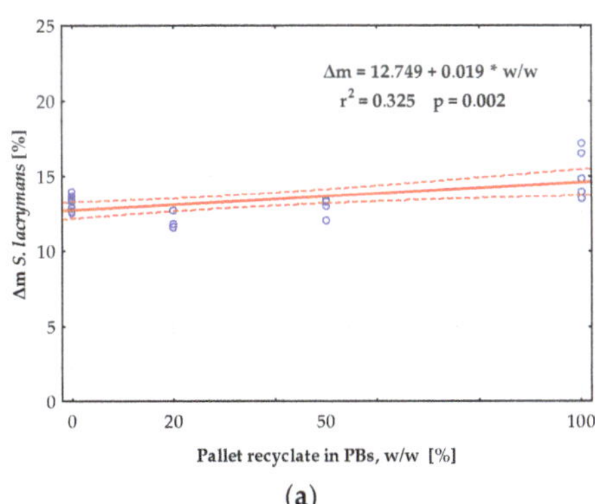

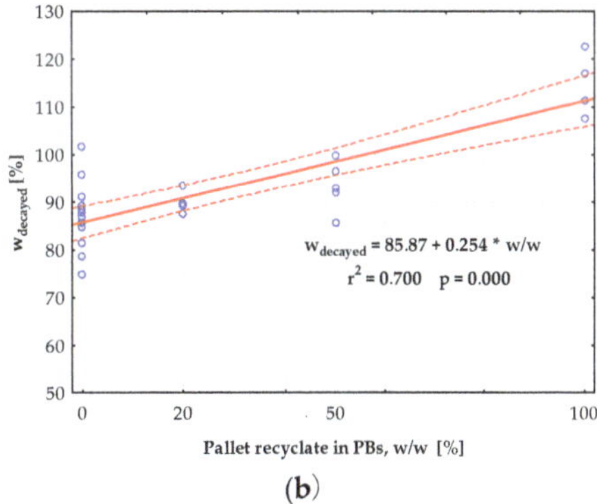

(**a**) (**b**)

Figure 9. Mass loss (Δm) (**a**) and moisture content ($w_{decayed}$) (**b**) of PBs containing different amount of particles from recycled spruce pallets (RSP) after attack by the brown-rot fungus *S. lacrymans*.

As mentioned above, spruce wood used for pallets can be deteriorated differently during storage and usage by several biological pests, such as decaying fungi, microscopic fungi, bacteria, etc. [48], at which the combined biological attack of spruce wood by more hts is usually more intensive; however, in Ref. [49], when spruce samples were primary intentionally damaged by one of the brown-rot fungi *Serpula lacrymans*, *Coniophora puteana* or *Gloeophyllum trabeum*, and then attacked again by other fungus of the aforementioned varieties, it was found that these brown-rot fungi are able to degrade the sound and primary rotten spruce wood with a comparable intensity. Some moulds (microscopic fungi) and bacteria can be useful for the subsequent wood decay caused by white-rot and brown-rot fungi; for example, *Bacilus* spp. bacteria disrupts pit membranes in cell walls of wood and thus helps the hyphae pass through tissues [50]; indeed, spruce samples primarily intentionally pretreated by the bacteria *Bacilus subtilis* had a slightly lower resistance to rot caused by the brown-rot fungus *S. lacrymans* comparing to sound spruce samples [51]. Generally can be stated, that a longer storage of spruce pallets or chips before their processing into PBs is undesirable.

The mould resistance of PBs evaluated after 7, 14, 21 and 28 days is summarized in Table 4. From the achieved results, it is evident that the kinetic and maximal rating values of the mould growth activity (MGA) on the top surface of PBs was not affected by particles from RSP; on the final (28th) day after inoculation, at least one half of PB surfaces was usually covered with hyphae of moulds, and the average MGA ranged from 3.67 to 4. Generally, surfaces of less durable and chemically unprotected woods and wooden composites are easily accessible to inhabitation by various moulds, mainly in a wet environment [52–54].

4. Conclusions

- The thickness swelling (TS) and water absorption (WA) values of the laboratory prepared particleboards (PBs)—based on particles from fresh spruce logs (FSL) and recycled spruce pallets (RSP)—were not affected by the particle type at all.
- On the contrary, the particles from RSP had a significantly negative effect on the mechanical properties of PBs—i.e., in connection with a decrease in the modulus of rupture (MOR) in bending up to 31.5% (from 14.6 to 10.0 MPa), the modulus of elasticity (MOE) in bending up to 23.1% (from 2616 to 2012 MPa), and the internal bond (IB) up to 22.8% (from 0.79 to 0.61 MPa).
- The particles from RSP had a significantly negative effect (but a maximum of 15.4%) on the decay resistance of PBs to the brown-rot fungus *Serpula lacrymans*. On the contrary, the mould resistance of PBs was not influenced by the type of spruce particles used.
- Generally, the manufacturing of PBs with the addition of recycled wood pallets is a very important issue from an economic and environmental point of view; however, the mechanical properties of PBs prepared from recycled wood pallets could be reduced, especially in those cases where in the used wood are damages caused by pests (fungi, insects, etc.), as well as at sorting undetected additives (biocides, paints, etc.), by which the wettability, adhesion and thus also the strength of the glued joints in the PBs is worsened.
- More representative conclusions on the optimization of PBs properties can be drawn only after more extensive laboratory and field research using several raw wood material sources, e.g., fresh logs, decayed elements from old buildings, recycled pallets and furniture, recycled modified wood, and of course, their mixtures.

Author Contributions: Conceptualization, J.I., L.R. and Z.V.; methodology, J.I., L.R. and Z.V.; software, J.I. and Z.V.; validation, J.I., L.R. and Z.V.; formal analysis, J.I. and L.R.; investigation, J.I., L.R. and Z.V.; resources, J.I., L.R. and Z.V.; data curation, J.I., L.R. and Z.V.; writing—original draft preparation, J.I., L.R. and Z.V.; writing—review and editing, J.I. and L.R.; visualization, J.I. and Z.V.; supervision, L.R.; project administration, J.I. and L.R.; funding acquisition, L.R. All authors have read and agreed to the published version of the manuscript.

Funding: This work was supported by the Slovak Research and Development Agency under the contract no. APVV-17-0583.

Institutional Review Board Statement: Not applicable.

Informed Consent Statement: Not applicable.

Data Availability Statement: The data presented in this study are available on request from the corresponding author. The data are not publicly available due to privacy restrictions.

Acknowledgments: We would like to thank to the Slovak Research and Development Agency under the contract no. APVV-17-0583 for support of the research published in this article. We also thank the companies Kronospan, s.r.o., Zvolen, Slovakia and Kronospan CR, spol. s.r.o., Jihlava, Czech Republic, for a providing parts of the materials for this research.

Conflicts of Interest: The authors declare no conflict of interest.

References

1. Lykidis, C.; Grigoriou, A. Hydrothermal recycling of waste and performance of the recycled wooden particleboards. *Waste Manag.* **2008**, *28*, 57–63. [CrossRef] [PubMed]
2. Ihnát, V.; Lübke, H.; Balberčák, J.; Kuňa, V. Size reduction downcycling of waste wood. Review. *Wood Res.* **2020**, *65*, 205–220. [CrossRef]
3. Zeng, Q.; Lu, Q.; Zhou, Y.; Chen, N.; Rao, J.; Fan, M. Circular development of recycled natural fibers from medium density fiberboard wastes. *J. Clean. Prod.* **2018**, *202*, 456–464. [CrossRef]
4. Michanickl, A. Recovery of fibers and particles from wood based products. In Proceedings of the No 7286: Wood and Paper in Building Applications, Forest Products Society, Madison, WI, USA, 9–10 September 1996; pp. 115–119.
5. Gaff, M.; Trgala, K.; Adamová, T. *Environmental Benefits of Using Recycled Wood in the Production of Wood-Based Panels*, 1st ed.; CZU: Prag, Czech Republic, 2018; p. 51.
6. Irle, M.; Privat, F.; Couret, L.; Belloncle, C.; Déroubaix, G.; Bonnin, E.; Cathala, B. Advanced recycling of post-consumer solid wood and MDF. *Wood Mater. Sci. Eng.* **2018**, *14*, 19–23. [CrossRef]
7. Lubke, H.; Ihnát, V.; Kuňa, V.; Balberčák, J. A multi-stage cascade use of wood composite boards. *Wood Res.* **2020**, *65*, 843–854. [CrossRef]
8. Altholz, V. Altholzverordnung. 15 August 2002. Available online: http://bundesrecht.juris.de/bundesrecht/altholzv/gesamt.pdf (accessed on 15 October 2021).
9. Khan, M.; Hussain, M.; Deviatkin, I.; Havukainen, J.; Horttanainen, M. Environ-mental impacts of wooden, plastic, and wood-polymer composite pallet: A life cycle as-sessment approach. *Int. J. Life Cycle Assess.* **2021**, *26*, 1607–1622. [CrossRef]
10. Leblanc, R. EPAL Pallet Production Increased in 2020 Despite COVID-19 Pandemic. Reusable Packaging News. Available online: https://packagingrevolution.net/epal-production-statistics/ (accessed on 13 April 2021).
11. Buehlmann, U.; Bumgardner, M.; Fluharty, T. Ban on landfilling of wooden pallets in North Carolina: An assessment of recycling and industry capacity. *J. Clean. Prod.* **2009**, *17*, 271–275. [CrossRef]
12. Roy, D.; Carrano, A.L.; Pazour, J.A.; Gupta, A. Cost-effective pallet management strategies. *Transp. Res. Part E Logist. Transp. Rev.* **2016**, *93*, 358–371. [CrossRef]
13. Deliverable: 1.1–Assessmentof Existing LTL Market, Products and Costs. Available online: https://ec.europa.eu/research/participants/documents/downloadPublic?documentIds=080166e5bf8af2a6&appId=PPGMS (accessed on 11 October 2021).
14. Gasol, C.M.; Farreny, R.; Gabarrell, X.; Rieradevall, J. Life cycle assessment comparison among different reuse intensities for industrial wooden containers. *Int. J. Life Cycle Assess.* **2008**, *13*, 421–431. [CrossRef]
15. Modern Materials Handling. Available online: https://www.mmh.com/article/annual_pallet_report_2021_short_supply_meets_high_demand=OK (accessed on 11 October 2021).
16. Hischier, R.; Althaus, H.-J.; Werner, F. Developments in wood and packaging materials life cycle inventories in ecoinvent (9 pp). *Int. J. Life Cycle Assess.* **2004**, *10*, 50–58. [CrossRef]
17. Iždinský, J.; Vidholdová, Z.; Reinprecht, L. Particleboards from recycled wood. *Forests* **2020**, *11*, 1166. [CrossRef]
18. STN 49 0110. *Wood—Compression Strength Limits Parallel to the Grain*; Slovak Office of Standards, Metrology and Testing: Brati-slava, Slovak, 1979.
19. EN 827. *Adhesives—Determination of Conventional Solids Content and Constant Mass Solids Content*; European Committee for Standardization: Brussels, Belgium, 2005.
20. EN ISO 2431. *Paints and Varnishes—Determination of Flow Time by Use of Flow Cups (Iso 2431:2019)*; European Committee for Standardization: Brussels, Belgium, 2019.
21. EN 1245. *Adhesives—Determination of pH*; European Committee for Standardization: Brussels, Belgium, 2011.
22. Iždinský, J.; Vidholdová, Z.; Reinprecht, L. Particleboards from recycled thermally modified wood. *Forests* **2021**, *12*, 1462. [CrossRef]

23. *EN 310. Wood-Based Panels—Determination of Modulus of Elasticity in Bending and of Bending Strength*; European Committee for Standardization: Brussels, Belgium, 1993.
24. *EN 317. Particleboards and Fibreboards—Determination of Swelling in Thickness after Immersion in Water*; European Committee for Standardization: Brussels, Belgium, 1993.
25. *STN 490164. Particle Boards—Determination of Water Absorption*; Slovak Office of Standards, Metrology and Testing: Bratislava, Slovak, 1980.
26. *EN 319. Particleboards and Fibreboards—Determination of Tensile Strength Perpendicular to the Plane of the Board*; European Committee for Standardization: Brussels, Belgium, 1993.
27. *EN 323. Wood-Based Panels—Determination of Density*; European Committee for Standardization: Brussels, Belgium, 1993.
28. *ENV 12038. Durability of Wood and Wood-Based Products—Wood-Based Panels—Method of Test for Determining the Resistance Against Wood-Destroying Basidiomycetes*; European Committee for Standardization: Brussels, Belgium, 2002.
29. *EN 15457. Paints and Varnishes—Laboratory Method for Testing the Efficacy of Film Preservatives in A Coating Against Fungi*; European Committee for Standardization: Brussels, Belgium, 2007.
30. *EN 322. Wood-Based Panels—Determination of Moisture Content*; European Committee for Standardization: Brussels, Belgium, 1993.
31. Mantanis, G.I.; Athanassiadou, E.T.; Barbu, M.C.; Wijnendaele, K. Adhesive systems used in the European particleboard, MDF and OSB industries. *Wood Mater. Sci. Eng.* **2018**, *13*, 104–116. [CrossRef]
32. Pizzi, A. Urea and melamine aminoresin adhesives. In *Handbook of Adhesive Technology*, 3rd ed.; Pizzi, A., Mittal, K.L., Eds.; CRC Press: Boca Raton, FL, USA, 2017; p. 38. [CrossRef]
33. *EN 312. Particleboards—Specifications*; European Committee for Standardization: Brussels, Belgium, 2010.
34. Müller, G.; Schöpper, C.; Vos, H.; Kharazipour, A.P.; Polle, A. FTIR-ATR spectroscopic analyses of changes in wood prop-erties during particle and fibreboard production of hard and softwood trees. *BioResources* **2009**, *4*, 49–71. [CrossRef]
35. Kuo, M.L.; McClelland, J.F.; Luo, S.; Chien, P.L.; Walker, R.D.; Hse, C.Y. Application of infrared photoacoustic spectroscopy for wood samples. *Wood Fiber Sci.* **1988**, *20*, 132–145.
36. Esteves, B.; Velez Marques, A.; Domingos, I.; Pereira, H. Chemical changes of heat treated Pine and Eucalypt wood monitored by FTIR. *Maderas-Cienc. Technol.* **2013**, *15*, 245–258. [CrossRef]
37. Popescu, M.-C.; Froidevaux, J.; Parviz Navi, P.; Popescu, C.-M. Structural modifications of *Tilia cordata* wood during heat treatment investigated by FT-IR and 2D IR correlation spectroscopy. *J. Mol. Struct.* **2013**, *1033*, 176–186. [CrossRef]
38. Akerholm, M.; Hinterstoisser, B.; Salmén, L. Characterization of the crystalline structure of cellulose using static and dynamic FT-IR spectroscopy. *Carbohydr. Res.* **2004**, *339*, 569–578. [CrossRef]
39. Popescu, M.-C.; Popescu, C.; Lisa, G.; Sakata, Y. Evaluation of morphological and chemical aspects of different wood species by spectroscopy and thermal methods. *J. Mol. Struct.* **2011**, *988*, 65–72. [CrossRef]
40. Wilcox, W.W. Review of literature on the effects of early stages of decay on wood strength. *Wood Fiber Sci.* **1978**, *9*, 252–257.
41. Rayner, A.D.M.; Boddy, L. *Fungal Decomposition of Wood: Its Biology and Ecology*, 1st ed.; John Wiley & Sons Ltd.: Chichester, UK, 1988; p. 587.
42. Azambuja, R.R.; Castro, V.G.; Trianoski, R.; Iwakiri, S. Recycling wood waste from construction and demolition to produce particleboards. *Maderas-Cienc. Technol.* **2018**, *20*, 681–690. [CrossRef]
43. Hameed, M.; Rönnols, E.; Bramryd, T. Particleboard based on wood waste material bonded by leftover cakes of rape oil. Part 1: The mechanical and physical properties of particleboard. *Holztechnologie* **2019**, *6*, 31–39.
44. Weber, C.; Iwakiri, S. Utilization of waste of plywood, MDF, and MDP for the production of particleboards. *Cienc. Florest.* **2015**, *25*, 405–413.
45. Nourbakhsh, A.; Ashori, A. Particleboard made from waste paper treated with maleic anhydride. *Waste Manag. Res.* **2009**, *28*, 51–55. [CrossRef]
46. Laskowska, A.; Maminski, M. The properties of particles produced from waste plywood by shredding in a single-shaft shredder. *Maderas-Cienc. Technol.* **2020**, *22*, 197–204. [CrossRef]
47. *EN 113. Durability of Wood and Wood-Based Products-Test Method Against Wood Destroying Basidiomycetes-Part 1: Assessment of Biocidal Efficacy of Wood Preservatives*; European Committee for Standardization: Brussels, Belgium, 2020.
48. Schmidt, O. *Wood and Tree Fungi–Biology, Damage, Protection, and Use*; Springer: Berlin/Heidelberg, Germany, 2006; p. 334.
49. Reinprecht, L.; Tiralová, Z. Susceptibility of the sound and the primary rotten wood to decay by selected brown-rot fungi. *Wood Res.* **2001**, *46*, 11–20.
50. Yildiz, S.; Canakci, S.; Yildiz, U.C.; Ozgenc, O.; Tomak, E.D. Improving of the impregnability of refractory spruce wood by Bacillus licheniformis pretreatment. *BioResources* **2011**, *7*, 565–577. [CrossRef]
51. Tiralová, Z.; Pánek, M.; Novák, S. Resistance of spruce wood pre-treated with the bacterium Bacillus subtilis and the moulds *Trichoderma viride* against selected wood-destroying fungi. *Acta Fac. Xylol. Zvolen* **2007**, *49*, 45–51.
52. Hosseinaei, O.; Wang, S.; Taylor, A.M.; Kim, J.W. Effect of hemicellulose extraction on water absorption and mold susceptibility of wood-plastic composites. *Int. Biodeterior. Biodegrad.* **2012**, *71*, 29–35. [CrossRef]
53. Arango, R.; Yang, V.; Lebow, S.; Lebow, P.; Wiemann, M.; Grejczyk, M.; DeWald, P. Variation in mold susceptibility among hardwood species under laboratory conditions. *Int. Biodeterior. Biodegrad.* **2020**, *154*, 105082. [CrossRef]
54. Aleinikovas, M.; Varnagirytė-Kabašinskienė, I.; Povilaitienė, A.; Šilinskas, B.; Škėma, M.; Beniušienė, L. Resistance of wood treated with iron compounds against wood-destroying decay and mould fungi. *Forests* **2021**, *12*, 645. [CrossRef]

 *forests*

Article

New Perspectives for LVL Manufacturing from Wood of Heterogeneous Quality—Part. 1: Veneer Mechanical Grading Based on Online Local Wood Fiber Orientation Measurement

Robin Duriot *, Guillaume Pot , Stéphane Girardon , Benjamin Roux, Bertrand Marcon, Joffrey Viguier and Louis Denaud

Arts et Metiers Institute of Technology, LaBoMaP, UBFC, HESAM Université, F-71250 Cluny, France; guillaume.pot@ensam.eu (G.P.); stephane.girardon@ensam.eu (S.G.); benjamin.roux@ensam.eu (B.R.); bertrand.marcon@ensam.eu (B.M.); joffrey.viguier@ensam.eu (J.V.); louis.denaud@ensam.eu (L.D.)
* Correspondence: robin.duriot@ensam.eu

Abstract: The grading of wood veneers according to their true mechanical potential is an important issue in the peeling industry. Unlike in the sawmilling industry, this activity does not currently estimate the local properties of production. The potential of the tracheid effect, which enables local fiber orientation measurement, has been widely documented for sawn products. A measuring instrument exploiting this technology and implemented on a peeling line was developed, enabling us to obtain the fiber orientation locally which, together with global density, allowed us to model the local elastic properties of each veneer. A sorting method using this data was developed and is presented here. It was applied to 286 veneers from several logs of French Douglas fir, and was compared to a widely used sorting method based on veneer appearance defects. The effectiveness of both grading approaches was quantified according to mechanical criteria. This study shows that the sorting method used (based on local fiber orientation and average density) allows for better theorical quality discrimination according to the mechanical potential. This article is the first in a series, with the overall aim of enhancing the use of heterogeneous wood veneers in the manufacturing of maximized-performance LVL by veneer grading and optimized positioning as well as material mechanical property modelization.

Keywords: laminated veneer lumber; veneer quality sorting; local fiber orientation; Douglas fir

Citation: Duriot, R.; Pot, G.; Girardon, S.; Roux, B.; Marcon, B.; Viguier, J.; Denaud, L. New Perspectives for LVL Manufacturing from Wood of Heterogeneous Quality—Part. 1: Veneer Mechanical Grading Based on Online Local Wood Fiber Orientation Measurement. *Forests* **2021**, *12*, 1264. https://doi.org/10.3390/f12091264

Academic Editors: Ladislav Reinprecht and Ján Iždinský

Received: 15 July 2021
Accepted: 13 September 2021
Published: 16 September 2021

1. Introduction

Laminated veneer lumber (LVL) is an engineered wood product made from veneers that is obtained by rotary peeling. It is used in many structural applications such as beams, columns, or panels due to its consistently high mechanical performance and improved dimensional stability in comparison with solid timber [1]. LVL-P is defined in the technical literature as a layup of 3 mm veneers laid in the same direction [1]. As a result, LVL-P is the LVL material that presents the best possible mechanical properties within the grain direction, and thus LVL-P beams are often produced using the highest strength grade of veneers to optimize the beam dimensions and provide good material efficiency. To obtain great dimensional stability relative to humidity conditions, LVL-P elements are usually slender beams with a thickness-to-width ratio of 1:8 at a maximum. They are mainly applied horizontally as flooring supports subjected to edgewise bending or as studs for wall structures [1].

In the peeling industry, the sorting of veneers to differentiate their quality is usually performed according to appearance criteria (knot size, slots, resin pockets, etc.) based on the EN 635-3 standard [2]. This classification is generally undertaken through the use of cameras integrated into the panel-making line. However, these appearance criteria present a low correlation with mechanical properties. Thus, some criteria should to be defined for

47

mechanical sorting purposes. Unlike in the peeling industry, in the sawmilling industry mechanical property assessment using grading machines (stress grading, vibrational tests) based on physical measurements has been developed. In recent scientific advances, two main technologies are used for better timber quality assessment: X-ray scanning, which provides local density through board thickness, and laser-dot scanning, which provides the local fiber orientation on board surfaces. With wood being a highly anisotropic material, the strength and stiffness properties are far better in the longitudinal fiber direction than in perpendicular. It appears that models based on fiber orientation measurement could provide greater improvements in bending strength prediction as compared to X-ray scanning [3]. This quantitative measurement of wood anisotropy may be the reason why fiber orientation-based models are better than local density (X-ray)-based models, which only indirectly estimate the wood properties from the local density (mainly clear wood density and knots) [4]. However, for species such as Douglas fir that are variable in terms of density and perturbated local fiber orientation, both should be considered.

The assessment of the mechanical potential of veneers by physical measures is therefore important for veneer grading, but the implementation of existing technologies from the sawmill to the peeling line is not simple. The X-ray scanning of veneers on a peeling line is expensive to implement, mainly because of the larger width of the veneers (several meters) as compared to sawn boards (up to 200 mm).

Conversely, laser-dot scanning is easier, cheaper, and safer to implement in a peeling line than X-ray scanning. Laser-dot scanning (also known as fiber-orientation scanning), is based on the light scattering of a laser dot projected on the wood surface, called the "tracheid effect" [5–7]. Laser dots on the wood surface scatter in quasi-elliptically shaped light spots according to fiber orientation. These ellipses can be acquired with standard cameras, and then the obtained pictures can be binarized. The contour of the binarized ellipses is fitted to an ellipse equation, and the angle between the major axis of the ellipse and the longitudinal direction of the board is defined. A great advantage of this method is that the measure can be performed on green wood veneers. Indeed, the high moisture content makes the ellipses larger and thus easier to measure with great accuracy, allowing the use of low and safe power lasers [7]. Moreover, green veneers are not subject to drying deformation in the 3D space, allowing for regular measurement in the LT plan. As a result, the in-line assessment device cost is cheaper and the protective fairing can be unfussy.

Much research using this technology has been performed recently: a similar modeling method of the mechanical properties of sawn timber was simultaneously developed by the LaBoMaP [8–10] and Linnaeus University of Vaxjö, Sweden [11–13]. Viguier et al. [14] successfully adapted this modeling method based on the fiber orientation scanning of small dimensions veneers that were scanned with a laser-dot scanner which usually worked with sawn timber. These authors showed that it was possible to model the flatwise bending stiffness of a LVL-P from the laser-dot scanning of the constituted veneers. Morover, they pointed out the fact that knowledge of the local fiber orientation is particularly important for homogeneous species such as beech. However, no study of the edgewise bending test was performed because of the small dimensions of the samples. Frayssinhes et al. [15] showed the first results of an online laser-dot scanner, the Local Online Orientation fiBer AnalyseR (LOOBAR), which is able to measure local fiber orientation directly during the rotary peeling process, and is thus not limited by the width of a sawmill type laser-dot scanner. These authors proved the ability of the device to measure the local fiber orientation and compared it to a fiber deviation model based on the distribution and size of knots, but no assessment of veneer mechanical properties was made.

The present paper aims to show the possible processing based on local fiber orientation measurements with the LOOBAR device in order to assess the mechanical properties of veneers and grade them in different strength classes. This article will first describe the LOOBAR constitution and its inner operations. Then, a method will be proposed that allows classes to be developed using physical parameters, combining the local fiber orientation and the average density. We propose testing the measurement and method on peeling and processing data from a very heterogeneous resource for which sorting is all the more necessary: the French large-diameter Douglas fir. The main objective is to compare this innovative sorting method to the sorting criteria method based on the visual observation of knots (like EN 635-3 standard [2]) at the scale of the veneers, and determine its consequences for LVL-P beams subjected to edgewise bending.

2. Materials and Methods

Table 1 shows is the nomenclature listing the main symbols used.

Table 1. Nomenclature.

List of Main Symbols			
$\overline{\rho_{veneer}}$	Averaged veneer density at 12% moisture content $(kg{\cdot}m^{-3})$	E_T	Sorting threshold in terms of modulus of elasticity (MPa)
$\theta(x,y)$	Local fiber orientation angle (°)	$\overline{E_{veneer}}$	Averaged local modulus of elasticity on all veneer surfaces (MPa)
$\overline{E_{beam}}(x,y)$	Average of $E_{ply,n}(x,y)$ of the 15 constitutive plies (MPa)	$\overline{E_x}(x)$	Stiffness profile: averaged local modulus of elasticity along the length of the veneer (MPa)
E_{eq}	Beam equivalent modulus of elasticity (MPa)	$E_x(x,y)$	Local modulus of elasticity of the veneer (MPa)
$EI_{eff}(x)$	Effective bending stiffness profile along the length of the beam $(N{\cdot}mm^2)$	$\overline{E_{x\,min}}$	Stiffness profile minimum value (MPa)
$EI_{eff\,min}$	Effective bending stiffness profile minimum value $(N{\cdot}mm^2)$	$\overline{H_{veneer}}$	Averaged normalized local Hankinson value on all veneer surfaces (dimensionless quantity)
$E_{ply,n}(x,y)$	Local modulus of elasticity of ply (MPa)	KSR	Knot surface ratio (d.q.)

2.1. Douglas-Fir Forest Stand and Peeling

Three different samples compose the Douglas-fir (*Pseudotsuga menziesii* (Mirb.) Franco) resource used for the experimental peeling/scanning (Table 2).

Table 2. Forest stands.

	Forest Stand 1	**Forest Stand 2**	**Forest Stand 3**
Locality	Sémelay (Nièvre, France)	Anost (Saône-et-Loire, France)	Cluny (Saône-et-Loire, France)
Average altitude	300–400 m	300–400 m	300–400 m
Cutting age	50–60 yo	50–60 yo	50–60 yo
Silviculture	Fourth thinned-out plot, non pruned wood	Clearcut, dynamic sylviculture, pruned wood	Classic (high forest)
Number of logs	4	4	7
Number of veneers	Sapwood: 68 Heartwood: 57	Sapwood: 8 Heartwood: 24	Sapwood: 47 Heartwood: 82
Average under bark trunk diameter	Min: 425 mm Max: 470 mm	Min: 410 mm Max: 445 mm	Min: 420 mm Max: 455 mm
Veneer average density	527 kg·m^{-3}	508 kg·m^{-3}	544 kg·m^{-3}

The peeling operation took place using the LaBoMaP rotary peeling machine in Cluny, France. All logs were preliminarily soaked in water at 50 °C before peeling during a minimum of 36 h, intending to ease the rotary peeling by increasing the material deformability [16]. The linear cutting speed was automatically controlled by the machine at 1.5 m·s^{-1} [17]. The rotary peeling operation was carried out without the use of a pressure bar since the lathe was not equipped with a roller pressure bar. This choice was made to avoid any problems coming from the Horner effect [18]. The peeling thickness was set to obtain dried veneers with a thickness of 3 mm. Scratcher knifes were set up, allowing the incision of the log on the surface prior to the cutting of the knife. This allowed a clean cut of veneer edges perpendicularly to the spindle axe. The distance between scratcher knives was adjusted to a nominal length of between 700 and 750 mm according to the length of each log. Concerning the cutting width of the veneers (perpendicular to the grain), it was adjusted according to the external aspect of the log (visible knottiness, radial cracks). A log that is radially cracked has a greater propensity to form discontinuities in the peeling ribbon. A short trimming length was chosen to maximize the number of uncracked veneers. The 2 chosen widths were 850 mm and 1000 mm. In total, the data coming from 286 veneers were treated in the following studies, coming from the peeling and online scanning of 12 Douglas-fir logs (see Table 2).

Heartwood was manually separated from sapwood through visual sorting based on veneer color. In Douglas fir, the color is highly differentiated between heartwood (salmon pink) and sapwood (white cream to yellow). This choice was made for numerical reasons in order to obtain enough sapwood veneer to compose the panels. The veneers were dried to a minimum of 9% moisture content in a drying stove for a minimum of 24 h. Moisture content was measured at the stove exit by double-weighing on control veneers [19]. The batches were then stored in open air to stabilize their moisture content with the ambient air (20 °C, 65% relative humidity).

2.2. Online Fiber Orientation Measurement Device: The LOOBAR

An innovative device was used to measure the fiber orientation of each veneer: the LOOBAR. It was developed within the LaBoMaP Wood Material and Machining (WMM) team and was integrated directly on the instrumented peeling line. The LOOBAR hardware, shown in Figure 1, is localized after the clipper of the peeling line, and thus it scans already clipped veneers (Figure 1a).

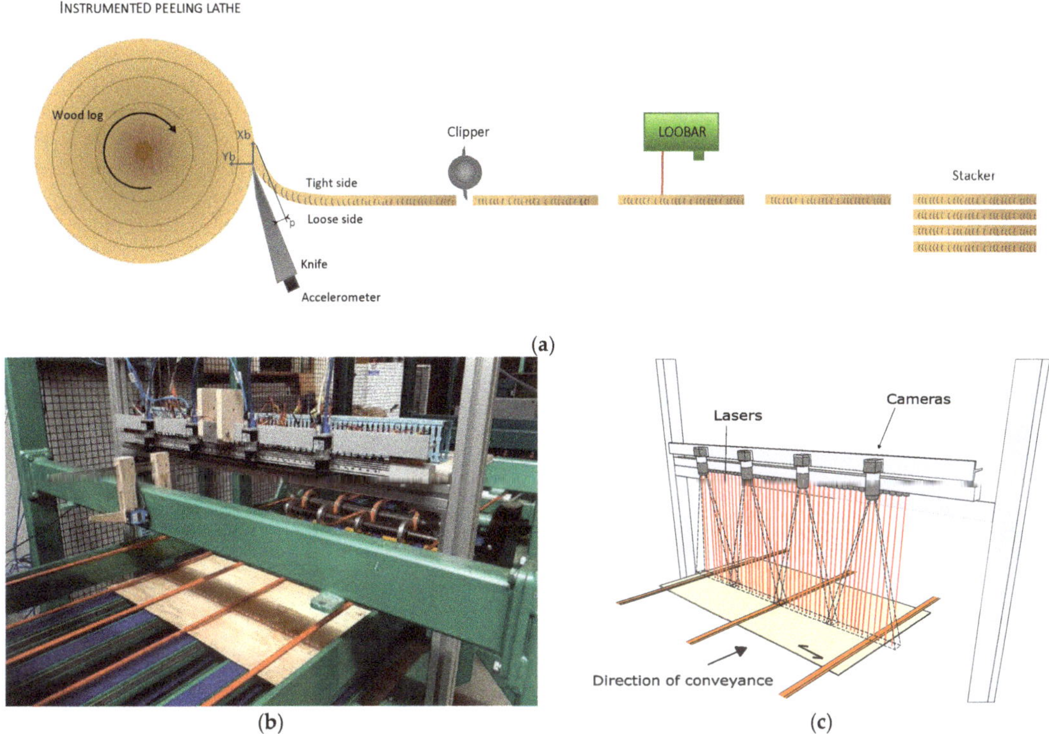

Figure 1. The LOOBAR, a veneer fiber orientation-measuring device integrated to the LaBoMaP peeling line: (**a**) schematic representation of LaBoMaP's instrumented rotary peeling line, (**b**) picture, (**c**) scheme.

Usually in a sawmill, industrial scanners use 4 rows of laser dots to scan the 4 sides of a timber board conveyed longitudinally at 1 time. Here, due to the low thickness of the veneers, the fiber orientation was considered to be the same across its thickness, justifying the measurement on only a single face.

As shown in Figure 1b,c, the LOOBAR is composed of:

- Fifty laser-dot modules with an output power of 5 mW each which project circular beams on veneers. They are positioned in line at a constant distance in the full 800 mm width of the peeling line, fixed on a support above it, and are thus perpendicular to the direction of the band and parallel to the main direction of the grain. Each laser-dot module is individually calibrated in order to provide a radiant flux of 0 to 5 mW (1 mW selected for Douglas fir) calibrated by a laser photodiode (Thorlab PM16-151). This power can be adapted according to the type of wood considered.
- Four cameras (Basler acA2440-75 μm), performing video acquisition of the reflected ellipses on the veneer surface. Their resolution is of 2048 pixels in the width of the line (or the main direction of the grain) and 120 pixels in the scrolling direction of the band (perpendicular to the main direction of the grain). The maximal acquisition rate is 1000 frames/s.

The measurement resolution in the main direction of the fibers (perpendicular to the conveyor) was about 16 mm, corresponding to the mean distance between each laser-dot module. The measurement resolution across the main fiber direction depends on the aquisition speed of the cameras and on peeling line veneer conveyor speed. For the linear cutting speed of 1.5 m·s^{-1} applied here to Douglas-fir veneer peeling, it was close to 3.5 mm. The 16 mm by 3.5 mm spatial resolutions obtained along and across the grain direction, respectively, can be compared to those of sawmill scanners, for which the corresponding resolutions are typically 1 mm in the length of the boards and 4 mm crosswise. A large difference in the resolution along main fiber direction according to the conveying direction was notable. In the main fiber direction, it was necessary to use a distance of 16 mm between the lasers so that the centers of the ellipses were not too close to avoid superimposing parts of ellipses. In the conveying direction, due to the movement of the veneers imposed by the linear speed of the line, ellipse shape information was obtained at different times, allowing for a much finer resolution (see Figure 2). Thus, in this direction, it was the acquisition frequency of the camera that drove the spacing of the ellipses.

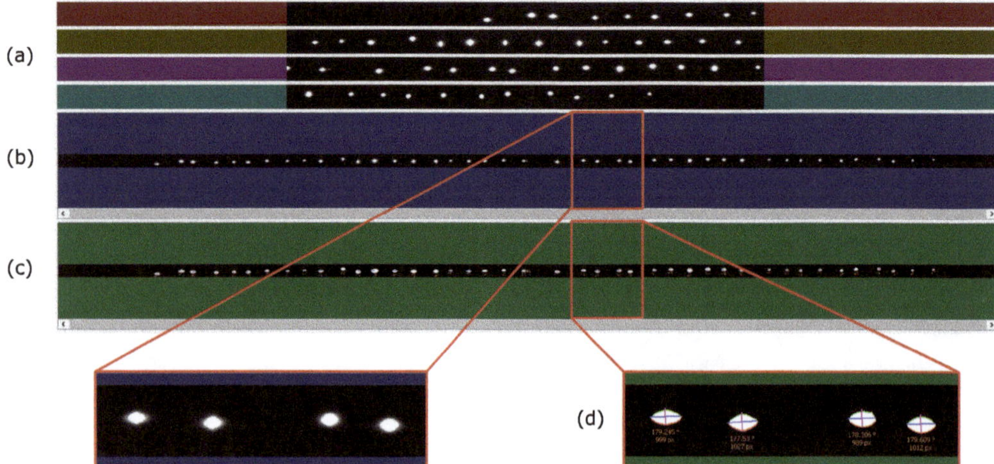

Figure 2. (**a**) Camera views to detect ellipses, (**b**) assembled view, (**c**) binarization step, (**d**) ellipse fitting.

Moreover, it should be noted that 3 belts ensured the conveying of the veneers along the line in the field of lasers and cameras (see Figure 1b,c). This therefore deprived the measurements of operating areas globally in the center and at the ends of the veneers. These laser points were therefore removed from the raw data.

A series of processing steps were required to convert the video raw data into a regular fiber orientation grid. The video information streams containing the raw data from the 4 cameras (Figure 2a) were recomposed into 1 as shown in Figure 2b. The processing of the superposition zones of the field of view of the cameras between them was carried out to guarantee the accuracy of the information over the entire width of the peeling line. A binarization of the images was then applied to detect the ellipse shapes (Figure 2c). In this study, the binarization threshold was adjusted for each peeled log as low as possible in order to obtain the largest possible ellipse areas without superimposition of the ellipses.

Veneer detection was performed by analyzing each of the images of the recorded ribbon: the increases without discontinuity in the average gray tone of the images corresponding to the passage of a light veneer over the dark background. Thus, each veneer was flagged by a beginning and an end in the temporal flow of the video, corresponding to a position and length in the ribbon once the veneer conveyor speed was taken into account. Processing was realized with the "FitEllipse" function of the OpenCV library [20] to enable the detection of ellipses and their orientations. This function calculates and provides the vertical and horizontal position of the center of the ellipse and the length of the large and small axis of the ellipse, as well as its angle with respect to the main direction of the fibers (Figure 2d). At the end of this first step of digital processing, 2 types of files were generated: a grayscale image for each veneer and a file containing information on the positioning and parameters of the ellipses.

After being cut by the clipper, veneers can slightly rotate during the conveyance. Thus, there was a second step of processing aiming at correcting the information by moving from the cameras coordinate system from the LOOBAR to the local coordinate system of the veneer. The detection of the cut edges and a conversion from px to mm allowed an angular correction of the fiber orientation values and a rotation of the grayscale image in order to visualize the veneer in its own local coordinate system (Figure 3a), with x_{green} being the position of a pixel in the longitudinal main direction of the veneer, and y_{green} being the position of a pixel in the tangential main direction of the veneer. This step allowe for the exploitation of the information even if the veneer coordinate system is not aligned with the coordinate system of the measurement apparatus.

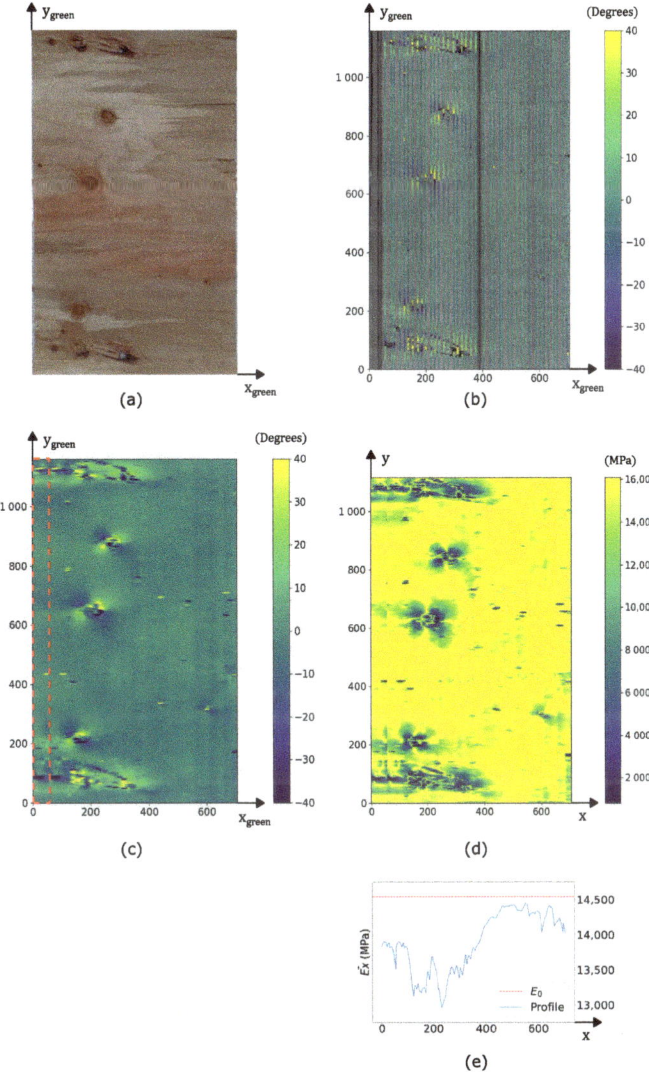

Figure 3. The veneer to fiber orientation grid: (**a**) veneer picture using a photo camera, (**b**) grayscale image and fiber orientation laser points, (**c**) interpolated local fiber orientation grid, (**d**) local elastic modulus grid, (**e**) stiffness profile.

Finally, the outputs of the LOOBAR scanning after these processing steps were as follows:

- x_{green}, the position of ellipses center along the grain direction in the veneer coordinate system (mm);
- y_{green}, the position of ellipse centers along the perpendicular axis to the grain direction in the veneer coordinate system (mm);
- θ, the angle (°);
- major and minor diameters (mm) that enable ratio calculation (dimensionless quantity); and
- the ellipse area (mm^2).

These first 3 bullet points are shown in Figure 3b.

2.3. From Local Fiber Orientation to Stiffness Regular Grid

From LOOBAR outputs it was possible to obtain a standard-property grid at 12% moisture content to conform to structural standard.

A local fiber orientation regular grid was extracted for each veneer, and then a linear interpolation of the measured raw data points corresponding to the centers of the ellipses over their entire surface (Figure 3c) was undertaken to obtain a regular grid with an X by Y resolution, as in [15]. The fiber deviation area was clearly visible in the vicinity of knots, with a positive or negative deviation that was almost symmetrical around them, as widely described in [15]. For the veneer ends with no angle data due to the presence of belts, a compensation algorithm was therefore developed for the problem of empty areas of mechanical properties in the manufacturing of numerical panels and beams (see Section 2.6). The adopted method consisted of replicating the contiguous zones in the empty zones and duplicating them (visible in red box of Figure 3c). Concerning the missing data due to the belts, as far as the central belt area was concerned, the applied interpolation enabled us to overcome the problem.

Using the regular grid of fiber orientation measurement of each veneer, the calculation of a local elastic modulus was performed in 2 steps following the method developed in Viguier's study [14]. First, the average density of the veneer was calculated. The mass of each veneer was weighed after drying at 9% with an accuracy of 0.1 g. For the calculation of the volume of the veneers, the nominal peeling thickness was used. The length and width at the green state were determined individually by manually cropping the grayscale images to the dimensions of each veneer ($\pm$5 mm). For width, the average shrinkage coefficient from the Tropix database [21] was applied according to the tangential orthotropic direction of the wood to adjust the volume to 9% moisture content. As a result, the veneer density was computed using Equation (1), including a 12% moisture content correction from the standard EN 384 [22].

$$\overline{\rho_{veneer}} = \frac{m_{veneer,9\%}}{L_{veneer,green}\, l_{veneer,9\%}\, e_{veneer,green}} \times [1 - 0.005(9\% - 12\%)] \tag{1}$$

where

- ρ_{veneer} is the veneer average density (kg·m^{-3}) at 12% moisture content;
- $m_{veneer,9\%}$ is the veneer global mass at 9% moisture content (kg);
- $L_{veneer,green}$ and $e_{veneer,green}$ are respectively the length and the thickness of the veneer at the green state (mm); and
- $l_{veneer,9\%}$ is the veneer width at 9% moisture content (mm).

The influence of density on E_0, the theoretical elastic of modulus for a straight grain, was computed according to Pollet [23], who studied clear Douglas-fir wood specimens by differentiating between juvenile and mature wood. Only the mature wood equation was used in this study because: (1) sapwood does not contain juvenile wood, and (2) the peeling operation leaves a core of wood at the pith with a diameter of 70 mm, which certainly contains almost exclusively juvenile wood concerning heartwood. Taking into account that the juvenile limit is usually observable for a cambial age of 15 years and by considering a mean radial growth of 5 mm for Douglas fir, a 75 mm diameter area from pith was obtained. Beyond this was the transition zone and mature wood. Equation (2), with 12% moisture content, is as follows:

$$E_0 = 36.605\,\overline{\rho_{veneer}} - 4242.4 \tag{2}$$

where

- E_0 is the theoretical elastic of modulus for straight grain (MPa); and
- ρ_{veneer} is the veneer average density (kg·m^{-3}) at 12% moisture content.

The effect of the local fiber orientation, noted as $\theta(x,y)$, on the local modulus of elasticity in the longitudinal main direction of the veneer, denominated as $E_x(x,y)$, was taken into account using Hankinson's formula (Equation (3)) [24], with parameters according to Wood HandBook data [25]:

$$E_x(x,y) = E_0\,\frac{k}{\sin^n(\theta(x,y)) + k \times \cos^n(\theta(x,y))} \tag{3}$$

where

- $\theta(x,y)$ is the local fiber orientation angle ($^\circ$) in the veneer coordinate system;
- k is the ratio between the perpendicular to the grain (E_0) and the parallel to the grain modulus of elasticity (E_{90}); it was taken as equal to 0.05, in accordance with [25], specifically for Douglas fir; and
- n is an empirical determined constant. It was taken as equal to 2, as recommended in [25] concerning the modulus of elasticity.

In order to be consistent with the actual 12% moisture content veneer density, the regular grid had to be resized to 12% moisture content dimensions. Thus, the average tangential shrinkage coefficient from the Tropix database [21] was applied for according to the orthotropic direction of the wood to adjust the veneer regular grid width. A regular grid of local modulus of elasticity at 12% moisture content for each veneer was finally obtained (Figure 3d). This consideration of shrinkage for veneer regular grid dimensions explains the change in the coordinate system from (x_{green}, y_{green}) to (x,y) and the reduction in height of Figure 3d according Figure 3c.

From $E_x(x,y)$ the averaged local modulus of elasticity on all veneer surface was calculated to provide a criterion to estimate the global quality of the veneer according Equation (4).

$$\overline{E_{veneer}} = \frac{\sum_{x=1}^{n_x}\sum_{y=1}^{n_y} E_x(x,y)}{n_x n_y} \tag{4}$$

where

- n_x is the number of pixels in the $\vec{x}$ direction; and
- n_y is the number of pixels in the $\vec{y}$ direction.

The normalized Hankinson mean value, which will be also evaluated in this study to provide a criterion on fiber orientation, is given as follows (Equation (5)):

$$\overline{H_{veneer}} = \frac{\sum_{x=1}^{n_x}\sum_{y=1}^{n_y} \frac{k}{\sin^n(\alpha(x,y)) + k \times \cos^n(\alpha(x,y))}}{n_x n_y} \tag{5}$$

This criterion is equal to 1 when the fiber orientation is perfecty longitudinal and tends to k when all fibers are perpendicular to the $\vec{x}$ direction in the veneer coordinate system.

2.4. From a Regular Grid to an Equivalent Longitudinal Stiffness Profile

Veneers are used in LVL beams where the veneer orientation is mostly all along the beam length. Thus, the longitunal modulus of elasticity is the main parameter for sorting the veneers. To sort them efficiently, an equivalent longitudinal stiffness profile was determined for each of the veneers.

The stiffness profile of each veneer, visible in Figure 3e, was obtained by averaging the modulus of elasticity values along the $\vec{y}$ direction, as calculated in Equation (6).

$$\overline{E_x}(x) = \frac{\sum_{y=1}^{n_y} E_x(x,y)}{n_y}$$

(6)

where n_y is is the number of pixels in the $\vec{y}$ direction.

Coarsely, the general elevation of the profile is given by the density, which affects the E_0 value (red dashed line in Figure 3e), and downward peaks are symptomatic of the presence of knots. These peaks are all the more important because knots are mostly aligned along the same x coordinate as they come from the same crown branch, as shown by the blue curve in the example of Figure 3e.

Tukey's HSD tests were performed to compare the criteria relative to each other.

2.5. Veneer Sorting and Grading

Two different ways to grade veneers are compared here based on appearance criteria on one hand, and based on average density and local fiber orientation on the other hand. This section describes each sorting and grading process.

2.5.1. Appearance Grading

Cropped grayscale images (Figure 4a) from the LOOBAR cameras were used to detect the knots of the veneers. Each veneer was processed manually using open source ImageJ image processing and analysis software [26]. Knots were identified by the grayscale nuance difference with clear wood and were then approximated by ellipses (Figure 4b).

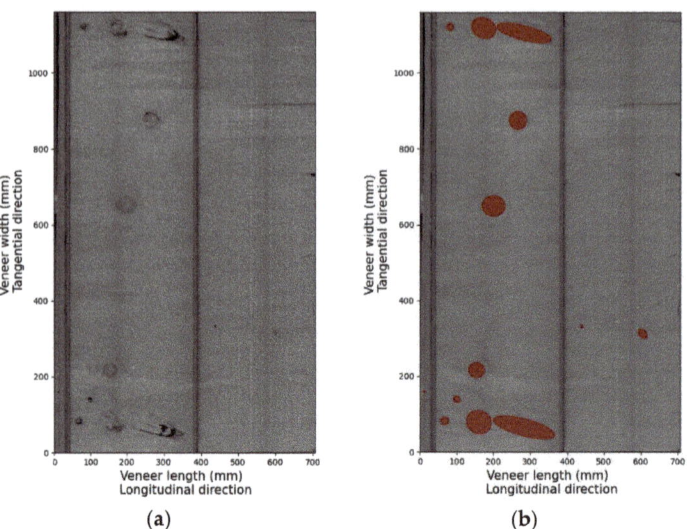

Figure 4. Veneer: (**a**) grayscale image, (**b**) manual knot detection.

Information regarding the centroïd coordinates (x_c, y_c) and the major and minor diameter (ϕ_{ma}, ϕ_{mi}), as well as the angle of the major diameter, was saved and stored. This detection of knots for each veneer allowed for the calculation of 2 criteria:

-   n_k, the number of knots per veneer; and

- *KSR* (knot surface ratio), defined as the ratio of cumulative knot areas to the total veneer area (Equation (7)),

$$KSR = \frac{\sum_{k=1}^{n_k} A_k}{A_v} \tag{7}$$

where

- A_k is the individual area of a knot (mm^2); and
- A_v is the area of the veneer (mm^2).

Veneers were sorted by considering the major diameter of the larger knot of each veneer. The classes of diameter were defined on the basis of EN 635-3 [2] criteria for adherent knots. After veneer drying, the knots were actually very brittle and close to be loose, but these criteria were better adapted to the population of diameters than the loose knots criteria of EN 635-3 which were designed for esthetic purposes. The classification conditions used are summarized in Table 3.

Table 3. Appearance grading classes from the EN 635-3 standard.

Categories of Characteristics	Appearance Grading Classes				
	E	I	II	III	IV
Very small knots	3/m^2 permitted	Permitted			
		Permitted up to an individual diameter of:			
Sound and adherent knots	Almost absent	15 mm as long as their cumulative diameter does not exceed 30 mm/m^2	50 mm	60 mm	Permitted

2.5.2. Modulus of Elasticity Profile Sorting

The aim was to no longer sort the veneers according to esthetic criteria but to use the composition of local fiber orientation and average density information. The stiffness profile of Equation (4) was used for sorting. In this study, a method was proposed that uses the minimum stiffness profile value of each veneer as the sorting criterion (Equation (8)).

$$\overline{E_x}_{min} = \min\left(\overline{E_x}(x)\right) \tag{8}$$

The value of this indicator is not impacted by the principle of replicated strips on the sides of veneers to fill in the lack of measurement. From a profile point of view, portions of the profile are replicated, which does not change the value of the general minimum.

Veneer classes were thus established. Two classes were envisaged: the high-quality class A and low-quality class B. Class A included veneers for which the elastic modulus profile minimum was superior than a given threshold over its entire length (green curve in Figure 5). Class B included veneers in which the elastic modulus profile minimum was lower than the threshold (red curve in Figure 5).

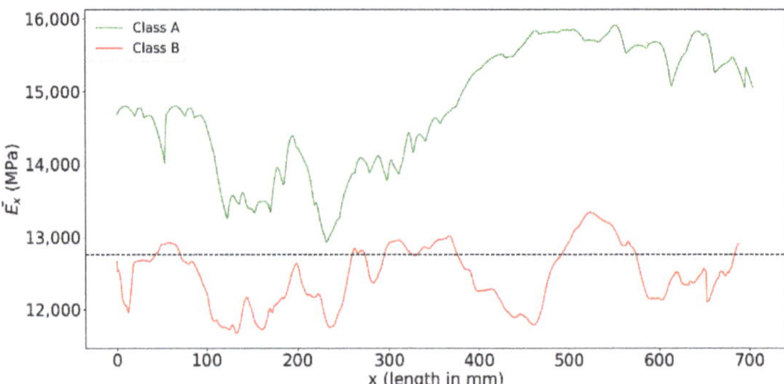

Figure 5. Example of sorting by the stiffness profile principle, with the dashed line being the sorting threshold.

A definition of a modulus of elasticity threshold between classes A and B is therefore necessary. Several thresholds can be defined according to the objectives of the manufacturer. Only 1 example has been retained for the present work. This threshold is the mean value of the minimum veneer stiffness profiles of the considered veneer population, defined by Equation (9). $\overline{E}_{xmin,v}$ is considered as the $\overline{E}_{xmin}$ value of a veneer v among a number of veneers n_v of a sample.

$$E_T = \frac{\sum_{v=1}^{n_v} \overline{E}_{xmin,v}}{n_v} \tag{9}$$

where n_v is the number of veneers of the considered sample.

2.6. LVL Panel Composition by Random Veneer Placement

In order to assess the mechanical properties of typical LVL-P beams made of a given class of graded veneers, an algorithm was developed based on the random placement of 60 actually measured veneers in the multi-ply material to simulate the manufacturing of a 15 ply LVL-P type panel. Assuming the thickness of each veneer as exactly 3 mm, this leads to a panel thickness of about 45 mm, which is a typical thickness for LVL-P [1]. The modeled panel length was 3000 mm (obtained by using 4 veneers per ply), so there was a total of 60 veneers per panel.

The algorithm steps are summarized in Figure 6. First, the input data were filled in: theoretical veneer dimensions (clipper width, peeling thickness, distance between scratcher knives), number of plies in the panel, the number of beams to be cut into the panel and their nominal dimensions. A sequence of stacking, listing all theoretical vacant positions where veneers would be assigned, was then created. To compose this sequence, a single theoretical length of veneer in the main grain direction (L_{theo}) was used. This corresponds to the distance between the scratcher knives which define the length of the veneers when peeling the logs. From 1 ply to the next a 1/5 length offset in the sequence of the theoretical length from the abscissa origin was performed. This offset from 1 ply to another, as for industrial classical products, aimed to avoid superimposing the veneer jointing areas and ensure the cohesion of the material. A 1/5 ratio is a good compromise between a too-small offset and a too-premature redundancy of the butt joint area.

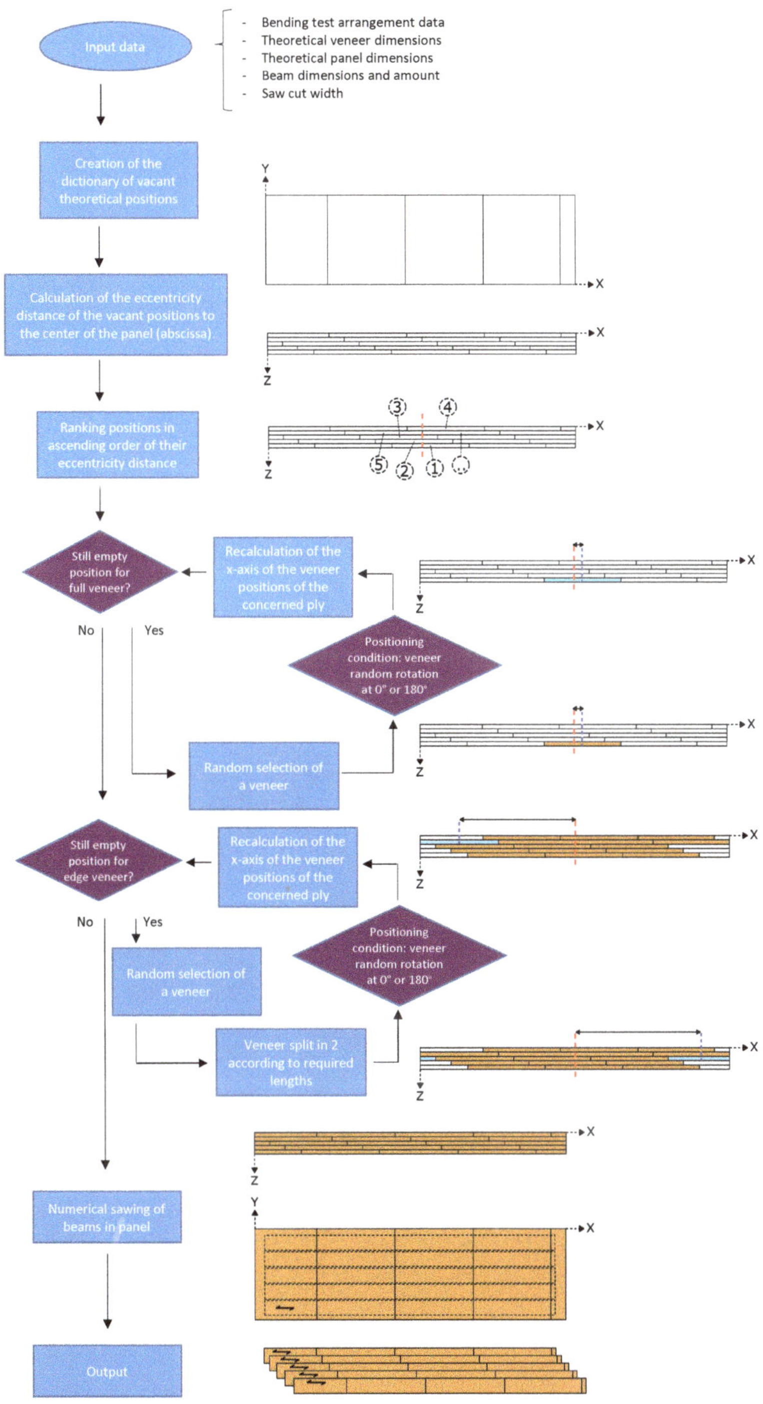

Figure 6. Random veneer placement algorithm.

In this study, the panel was composed of 60 veneers composed of 72 positions considering veneers at the extremities that were cut in half. In the initial state, the following theoretical information was available for each of the vacant veneer positions (for i from 1 to 72):

- $\vec{X}$: the axis parallel to the main orientation of the fibres in the panel;
- $X_{i,min,theo}$: in the panel system of the axis, this is the minimum abscissa of the theoretical position of veneer position number I;
- $X_{i,max,theo}$: in the panel system of the axis, this is the maximum abscissa of the theoretical position of veneer position number i. For each vacant full veneer position, there is $L_{theo} = X_{i,max,theo} - X_{i,min,theo}$;
- The ply number (between 1 and 15 for these 15 ply panels); and
- $X_{i,eccentricity,theo}$: The eccentricity distance. This is the distance between the abscissa of the centre of the panel and the centre abscissa of the theoretical position of the veneer position number i (Equation (10)).

$$X_{i,eccentricity,theo} = \left| -X_{i,\,centerpanel,\,theo} + \left(X_{i,min,theo} + \frac{X_{i,max,theo} - X_{i,min,theo}}{2} \right) \right| \qquad (10)$$

Each of the positions in the panel was initially considered as available. All vacant positions were first classified in ascending order according to their eccentricity distance. The first vacant position to be assigned was that with the lowest absolute value of the eccentricity distance. This method allowed us to start the algorithm by using full-length veneers, whereas at the ends of the panels many veneers had to be cut (cf. Figure 6). At each loop iteration a veneer was randomly selected from the list and placed in a vacant position. A random choice between placing the veneer at $0°$ or $180°$ was made at this step. Each time a veneer was allocated to a position, the theoretical abscissa of the vacant positions was recalculated from its real length, L_{veneer}. The difference between the theoretical length of the vacant position L_{theo} and the actual proper length of the veneer L_{veneer} assigned to it creates a offset, requiring the offset of the following veneer vacant positions in the concerned ply for each loop. Once all the positions requiring full-length veneers were assigned, the end veneers were in turn completed. For simplicity, when a veneer attributed to a ply was cut in 2 parts as required, each of its 2 parts completed both ends of the ply. Again, for each part of veneer the condition of $0°/180°$ for positioning was applied. Once all veneers had been assigned and therefore all plies of the material panel were filled, numerical sawing of the beams in the panel was performed according to the parameterized dimensions.

In this study, for each set of 60 veneers, 150 random panels were simulated. In order to evaluate their bending stiffness, each of them was virtually cut into 5 LVL-P beams of 145 mm in width. This width was chosen to cut beams with the minimum of 18 times the beam height between the lower support ratio from manufactured panels.

2.7. Mechanical Properties of Randomly Composed LVL Panels and Beams

This section details the calculation of the criterion used to quantify the mechanical bending performance of the beams randomly generated in bending.

In a first step, the $E_{ply}(x,y)$ value of each ply composing the beam was averaged along the $\vec{z}$ direction to obtain the grid $\overline{E_{beam}}(x, y)$ of the modulus of elasticity along the $\vec{x}$ and $\vec{y}$ direction of each beam according to Equation (11):

$$\overline{E_{beam}}(x, y) = \frac{\sum_{n=1}^{n_{plies}} E_{ply,n}(x, y)}{n_{plies}} \qquad (11)$$

where n_{plies} is the total number of plies in the $\vec{z}$ direction.

An effective bending stiffness, EI_{eff}, was calculated for each section of 1 pixel of height in the $\vec{y}$ direction along the $\vec{x}$ direction of the LVL beams in accordance with Equation (12):

$$EI_{eff}(x) = \sum_{y=1}^{n_y} \left(\overline{E_{beam}}(x,y) I_{gz,\,\Delta y} + \overline{E_{beam}}(x,y) A_{\Delta y} d_{\Delta y}(x)^2 \right) \tag{12}$$

with

$$I_{gz,\,\Delta y} = \frac{L_{beam} \times (\Delta_y)^2}{12}$$
$$A_{\Delta y} = L_{beam} \times \Delta_y$$

where

- n_y is the number of pixels in the $\vec{y}$ direction;
- $I_{gz,\Delta y}$ is the second moment of area of the section of Δ_y height at a given x position;
- $A_{\Delta y}$ is the area of the section of Δ_y height at a given x position; and
- $d_{\Delta y}$ is the distance from the neutral fiber of each element. This was calculated at a given x position and took into account the modulus of elasticity variation in the section, as explained in [14].

An equivalent modulus of elasticity, E_{eq}, was calculated according Equation (13), to establish a bending criterion for ranking the beams over their entire length.

$$E_{eq} = \frac{n_x}{I_{gz} \sum_{x=1}^{n_x} \frac{1}{EI_{eff}(x)}} \tag{13}$$

E_{eq} was chosen as the performance indicator for the bending beam because it considers the entire length of the beam in its calculation without considering a specific bending solicitation. Its calculation considers each elemental effective bending stifness at an x given position as being in a series along the entire length of the beam. The calculation of the equivalent module follows the same principle as that of an equivalent rigidity of several springs in a series. $\overline{E_{eq}}$ is considered as the E_{eq} mean value of beams in a given sample.

3. Results

3.1. A Comparison of Two Sorting Methods

3.1.1. Appearance Sorting

In total, 286 studied veneers were sorted according to the criteria considered in Section 2.5.1 from the EN 635-3 standard [2] relative to knottiness. A differentiation between heartwood and sapwood was also made to highlight distinctions regarding the impact of their morphological and physico-mechanical properties on the modeled performance of LVL products. Figure 7 shows the results of the number of veneers in each appearance class.

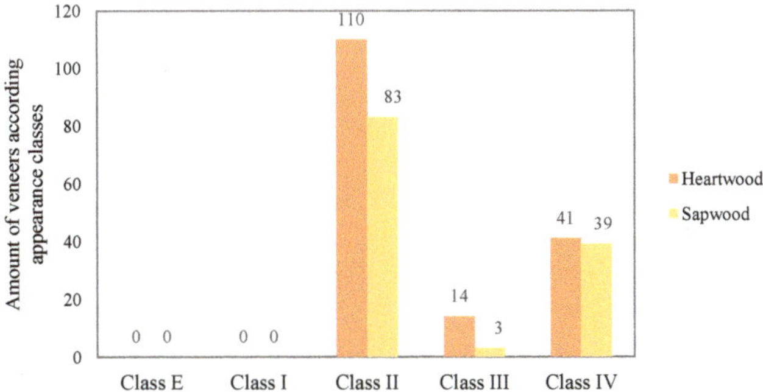

Figure 7. Distribution of veneers by appearance sorting class.

In the veneer sample there were no veneers in Classes E and I, as the criteria for knot size and concentration are very restrictive. This is very typical of veneers from large Douglas-fir trees for which large knots are very common. Class III is very low fill (14 veneers for heartwood, 3 for sapwood, or 5.9% of total amount of veneers). This was due to the standard that permits individual diameters of knots up to 60 mm, while in Class II they are permitted up to 50 mm. In other words, the largest knot of the veneer must be between 50 mm and 60 mm. This criterion was very restrictive as compared to that of Class IV, which allowed diameters of 60 mm and above. With regard to the proportion of heartwood and sapwood in Classes II and IV, it was not possible to draw conclusions regarding the relationship between the distance to the pith of each type of wood and knot sizes because the numbers of total veneers of sapwood and heartwood were different. Given the small number of veneers contained in Class III, the data from these veneers were regrouped with Class IV for the sake of simplicity.

Tukey's HSD test was performed to compare both the appearance criteria and computed criteria with the LOOBAR (Table 4). Populations submitted to an HSD test (Tukey's method) are signaled in the following tables by a lowercase letter. In order to facilitate the reading, a color code for each line was added to highlight the significantly different values (p-value < 0.05) line by line: green for the highest value, red for the lowest value.

Table 4. Data from appearance classes.

	Units	Heartwood Mean Value (CoV%)		Sapwood Mean Value (CoV%)	
		Class II	Class III/IV	Class II	Class III/IV
Amount of knots	/	20 (47) a	16 (39) b	9 (40) c	7 (32) c
KSR	%	0.58 (59) b	1.79 (25) a	0.28 (47) c	1.71 (45) a
$\overline{\rho_{veneer}}$	kg/m^3	503 (4.8) d	531 (5.2) c	553 (4.5) b	571 (5.4) a
$\overline{H_{veneer}}$	d.q.	0.936 (3.7) b	0.887 (3.9) c	0.950 (3.5) a	0.896 (4.8) c
$\overline{E_{veneer}}$	MPa	13,276 (7.1) b	13,492 (7.9) b	15,207 (6.7) a	14,930 (8.8) a

Values followed by a different letter within a column are statistically different at p-value = 5% (ANOVA and Tukey's HSD test).

First of all, we can make observations based on knottiness. The KSR value was higher in the lower-quality Class III/IV than in Class II for both heartwood and sapwood, is consistent with the principle of appearance grading. It appears that there were fewer knots in Class III/IV than in Class II for heartwood. For sapwood, the HSD test showed that there were no significant difference in the number of knots between Class II and Class III/IV veneers. There were fewer knots in sapwood than in heartwood, whereas the KSR values in Class III/IV were similar, showing that sapwood knots were much bigger that heatwood knots, which is consistent with the increasing diameter of a branch along with the growth of the tree.

Sapwood density was higher than that of heartwood, which is a classical effect due to density increase with the distance to the pith within the tree. More surprisingly, for both heartwood or sapwood the average veneer density of Class III/IV was significantly higher than that of Class II. This result may be explained by two effects:

1. The appearance sorting method select veneers with larger knots for the lower-quality Class III/IV. Veneers located far from the pith of the tree should arise (otherwise branches are not large enough). This is wood with a higher density, for the same reason as explained before.
2. Knots are denser than in clear wood and higher KSR values can be observed for class III/IV.

For both sapwood and heartwood, the average standardized Hankinson criterion $\overline{H_{veneer}}$ values were higher for the higher-quality Class II. This criterion is calculated from the measure of local fiber orientation, the variation of which is mainly due to the presence of knots.

The sapwood $\overline{E_{veneer}}$ was significantly higher than that of heartwood, but when looking at the classes from each type of wood (+12.6% between $\overline{E_{veneer}}$ of each type of wood), there was no significant difference according to the HSD test. This can be explained by a compensation phenomenon between the fiber orientation information and density, which results in a lack of real variation of this parameter based on appearance sorting classes.

It can therefore be concluded that, in terms of $\overline{E_{veneer}}$ for these studied batches, each appearance sorting class has similarly variable veneers. Sorting based solely on appearance may therefore not be the most efficient method because it does not take into account important parameters such as density, and it does not differentiate veneers with a few large knots from veneers with many smaller knots. Other sorting methods should therefore be considered.

3.1.2. Stiffness Profile Sorting

The stiffness profiles of all the veneers colored depending on their class were calculated for sapwood and heartwood according to the minimum method, as shown in Figure 8. For the sake of clarity, only one profile of each class is shown in bold and dark colour, while the remaining veneers are shown with a finer width and a clearer color.

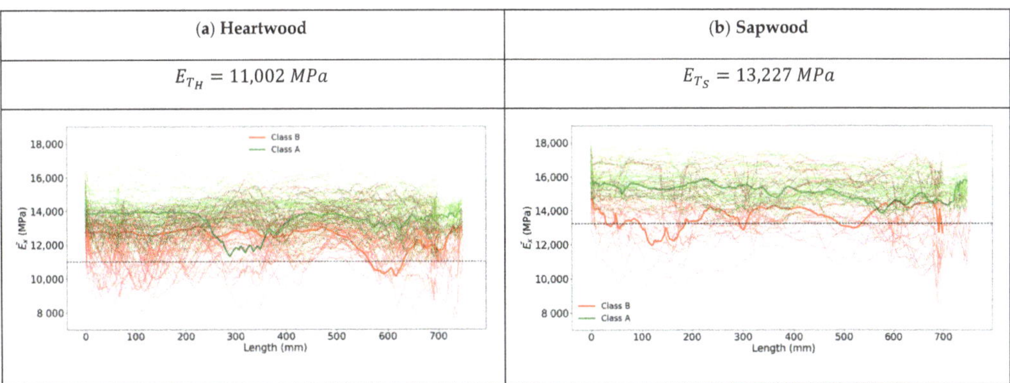

Figure 8. Veneer sorting by stiffness profile according to several methods (12% MC): (**a**) heartwood, (**b**) sapwood.

When performing appearance sorting, a single limiting knot is used to classify the veneer. By analogy, sorting by the limiting minimum value of the profile retains the same strategy. It is logical to note that the threshold was higher for sapwood than heartwood, since it was calculated from populations. Figure 9 shows veneer populations in Class A and Class B, including the proportion of each appearance sorting class. Since the threshold was the average of the individual minimum values of the veneer stiffness profiles, there was a logical balanced distribution of veneers between Classes A and B.

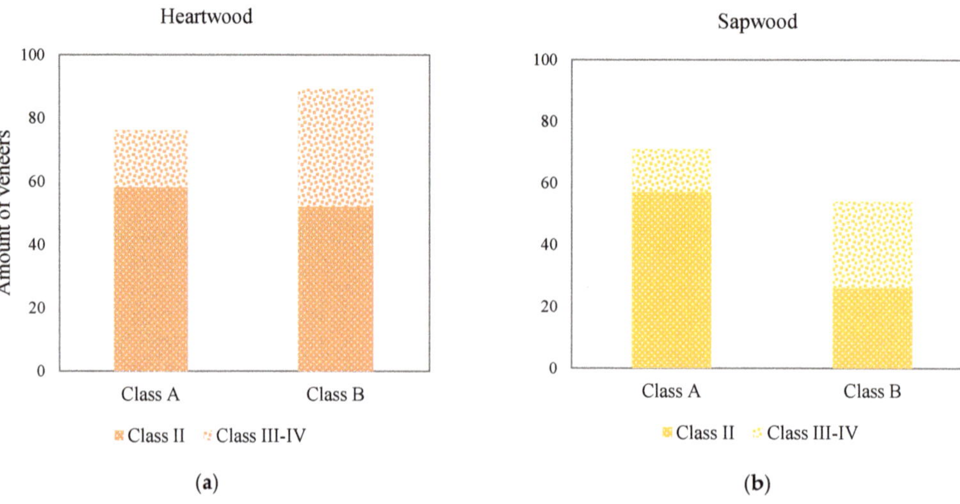

Figure 9. Distribution of veneers according to their sorting class by stiffness profile: (**a**) heartwood, (**b**) sapwood.

For heartwood, 76 veneers were distributed in Class A and 89 in Class B. For sapwood, 71 were in Class A and 54 in Class B. Similar proportions of appearance in Class II veneers were found in Classes A and B for heartwood (76.3%) and sapwood (80.3%). This implies, therefore, that 23.7% and 19.7% of appearance-sorted low-quality Class III/IV heartwood and sapwood veneers were incorporated into this new high-quality Class A. Conversely, 58.4% heartwood and 48.1% sapwood in appearance-sorted Class II veneers represented Class B. These results mean either that the distribution of the knots of these Class B veneers was concentrated in verticils, resulting in a minimum localized value, or that their density was in the low band. According to this approach (computed $\overline{E_x}(x)$) and chosen threshold, the appearance quality sorting is a very poor criterion for the grading of veneers from this Douglas-fir resource from a mechanical application perspective. Table 5 shows data regarding stiffness profile classes. This should be compared with Table 4.

Table 5. Data from stiffness profile classes.

		Heartwood Mean Value (CoV%)		Sapwood Mean Value (CoV%)	
	Units	Class A	Class B	Class A	Class B
Amount of knots	/	16 (43) b	21 (45) a	8 (36) c	8 (41) c
KSR	%	0.70 (82) b	1.22 (56) a	0.45 (98) c	1.16 (85) a
$\overline{\rho_{veneer}}$	kg/m^3	523 (5.3) c	504 (5.2) d	563 (4.4) a	553 (5.7) b
$\overline{H_{veneer}}$	d.q.	0.941 (3.1) a	0.902 (4.7) b	0.953 (2.5) a	0.905 (5.7) b
$\overline{E_{veneer}}$	MPa	13,996 (5.8) c	12,784 (6.0) d	15,575 (5.0) a	14,479 (8.3) b

Values followed by a different letter within a column are statistically different at *p*-value = 5% (ANOVA and Tukey HSD test).

It can be noted that the trends were no longer the same as those observed with appearance sorting. The upper class by stiffness profile (Class A) was now denser than the lower class (Class B) (+3.8% for heartwood, +1.8% for sapwood), whereas the reverse was true for appearance sorting classes (−5.3% and −3.1%, respectively). For the normalized Hankinson mean value $\overline{H_{veneer}}$ criterion, Class A was always higher than Class B (+4.3% and +5.3%), but the differences were now logically less important than for appearance sorting (+5.5% and +6.0%). This trend was also logically followed for KSR. For the averaged local modulus of elasticity mean value $\overline{E_{veneer}}$, there were now significant differences between Class A and Class B. Class A had veneers with higher average values of +9.5% and +7.6%, respectively, with regard to Class B veneers. When sapwood veneers presented a higher $\overline{E_{veneer}}$ than heartwood, the Class A heartwood $\overline{E_{veneer}}$ value was 13,996 MPa, slightly higher than the Class II appearance value of 13,276 MPa.

3.2. Comparison of LVL Mechanical Performance According to Sorting Method and Classes

Figure 10a,b presents an example of a regular grid of the local modulus of elasticity averaged over the 15 plies ($\overline{E_{beam}}(x,y)$) of the LVL beam randomly generated from veneers classified in Class A that are relatively well homogenized.

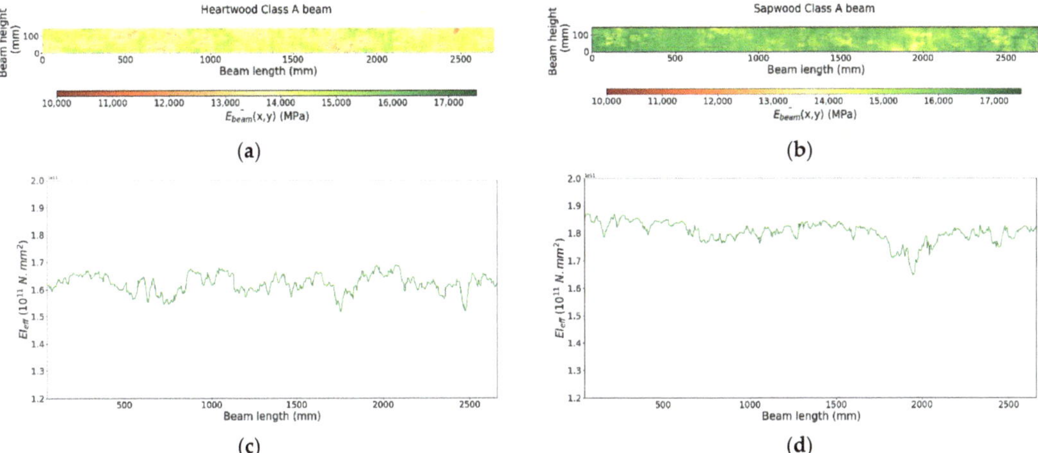

Figure 10. Random beams: (**a**) heartwood $\overline{E_{beam}}(x,y)$, (**b**) sapwood $\overline{E_{beam}}(x,y)$, (**c**) heartwood EIeff(x) profile, (**d**) sapwood EIeff(x) profile.

Figure 10c,d shows an example of an effective stiffness profile of beams. Because of its overall constant appearance, this makes it possible to observe the homogeneity in terms of the EI_{eff} profile of the random placement of heterogenous veneers in the material.

In total, 150 random panels from each class were modeled, in which 5 beams were digitally cut. When a class had fewer than 60 veneers, some were doubled representatively of the forest plots and logs which composed the batches. These beams were then tested by the analytical calculation model. The aim was therefore to be able to compare the performance of beams composed of veneers from classes. $\overline{E_{eq}}$ was calculated for each class. Figure 11 shows results for each class.

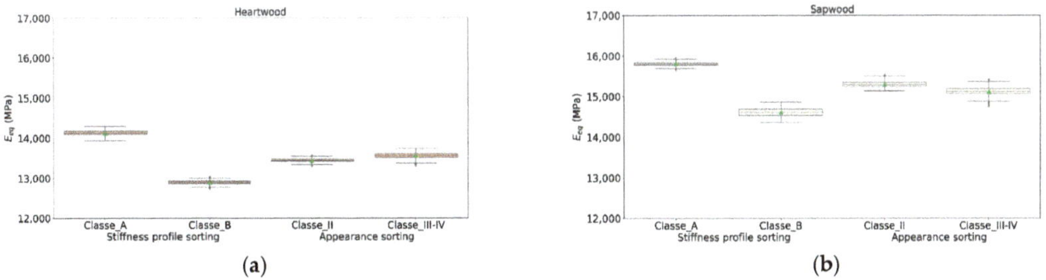

Figure 11. E_{eq} criteria according classes (12% MC): (**a**) heartwood, (**b**) sapwood.

In the case of profile sorting, the E_{eq} results of panel beams made up of Class A veneers were higher than those of Class B beams. The consideration of fiber orientation and density in the grading methodology appears to be more effective because it reveals a better discrimination of qualities between upper and lower classes than the actual standard of appearance sorting.

To ensure the reliability of the $\overline{E_{eq}}$ results, 150 control random panels were generated for each class. It appears that for the results presented, the maximum relative difference between sample presented and control samples $\overline{E_{eq}}$ was always less than 0.2%.

For sapwood, it is noted that the beams from the Class A random panels, which is the top-graded class, had an $\overline{E_{eq}}$ that was 8.2% higher than that of the Class B random beams (15,805 MPa vs. 14,610 MPa). The difference between appearance-sorted Classes II and III/IV averaged 1.2% (15,301 MPa vs. 15,119 MPa). For heartwood, there was a 9.6% difference (14,126 MPa vs. 12,894 MPa) between Class A and Class B beams. By comparison, the difference between appearance-sorted classes II and III/IV was on average equivalent (13,442 MPa vs. 13,555 MPa). It can therefore be concluded that the sorting by stiffness profile is much more selective and nicely dissociates the product qualities from the local modulus.

It is interesting to put these results into perspective with those of Tables 4 and 5. There was a very minor gap between the $\overline{E_{veneer}}$ of the veneers and the E_{eq} of the beams randomly modeled for each class, whether they were from visual sorting or profile sorting (maximum 1.4% relative gap).

4. Discussion

This article provides the first results of veneer sorting based on both fiber orientation and density considerations.

It was shown that veneer sorting by stiffness profile computed from fiber orientation measurement and density mean values theoretically leads to more efficient grading than appearance grading based on knottiness to obtain a higher stiffness. It is important to remember that this conclusion is based on modeling results that allowed us to obtain a large number of results, which would not be attainable by experimental tests (a total of 600 panels or 3000 beams were generated). However, the conclusion should be trusted because the mechanical model principles relies on a method which has been successfully applied to sawn timber [10,27]. Furthermore, in the second part of this series of articles [28], the present modeling method is successfully compared to experimental measurement with some very different testing configurations.

It should be noted that only one particular method to separate the batch of veneers in two classes has been tried, but many others could have been considered. In addition, the limited studied veneer population, especially by practicing a heartwood/sapwood differentiation, does not allow for the creation of more than two classes. With more veneers and even more varied forest stands, more classes would be possible. This could take place for instance by using a percentile of stiffness instead of the mean value or global value, which can rely on strength classes (EN 338 standard [29]). In this article, the "minimum method", based on the analogy of appearance grading, may not be sufficient. Use of the stiffness profile minimum and the MOE mean value of veeners computed into a single multi-criteria score may be better.

This sorting method, based on the modeling of local mechanical properties by density and local fiber orientation measurement, offers very interesting potential for the peeling industry. Since the thresholds are calculated from measured veneers, a two-step procedure for implementing this sorting method could be proposed. An initial phase of representative resource measurement for considered species could be performed to calculate the thresholds. Sorting could then be implemented, and the calculation of thresholds refined by the continuously generated data. The moisture-sorting process existing in the industry, which allows for the discrimination of heartwood and sapwood, can be seen has a form of quality sorting in itself, which can lead to dissociated application fields. Heartwood, due to its durability, can be used in outdoor structures such as bridges. Sapwood, with better intrinsic properties due in part to its higher density, it can be used in heavy-stressed indoor or outdoor construction with the application of protective products.

Author Contributions: Conceptualization, R.D.; methodology, R.D.; software, B.R. and R.D.; validation, G.P., S.G. and L.D.; formal analysis, R.D.; investigation, R.D., G.P., S.G. and L.D.; resources, B.R., B.M. and J.V.; data curation, R.D.; writing—original draft preparation, R.D.; writing—review and editing, R.D., G.P., S.G., L.D., B.R., J.V. and B.M.; visualization, R.D.; supervision, G.P., S.G. and L.D.; project administration, L.D. and G.P.; funding acquisition, G.P., B.M., B.R. and L.D. All authors have read and agreed to the published version of the manuscript.

Funding: This study was funded by the region of Burgundy Franche-Comté and the TreeTrace project subsidized by ANR-17-CE10-0016-03. This study was performed thanks to the partnership built by BOPLI, a shared public-private laboratory built between the Bourgogne Franche-Comté region, LaBoMaP, and the BRUGERE company.

Data Availability Statement: The data presented in this study are available on request from the corresponding author. The data are not publicly available due to industrial application confidentiality.

Acknowledgments: Thanks to Jean-Claude Butaud for the experimental task preparation and execution. Thanks to Remy Frayssinhes in the peeling task.

Conflicts of Interest: The authors declare no conflict of interest. The funders had no role in the design of the study; in the collection, analyses, or interpretation of data; in the writing of the manuscript, or in the decision to publish the results.

References

1. *LVL Handbook Europe*; Federation of the Finnish Woodworking Industries: Helsinki, Finland, 2019.
2. NF EN 635-3. Plywood. Classification by Surface Appearance. Part 3: Softwood. 1 July 1995. Available online: https://sagaweb-afnor-org.rp1.ensam.eu/fr-FR/sw/consultation/notice/1244973?recordfromsearch=True (accessed on 29 October 2020).
3. Faydi, Y.; Viguier, J.; Pot, G.; Daval, V.; Collet, R.; Bleron, L.; Brancheriau, L. Modélisation des propriétés mécaniques du bois à partir de la mesure de la pente de fil. In Proceedings of the 22e Congrès Français de Mécanique, Lyon, France, 24–28 August 2015.
4. Viguier, J.; Marcon, B.; Girardon, S.; Denaud, L. Effect of Forestry Management and Veneer Defects Identified by X-ray Analysis on Mechanical Properties of Laminated Veneer Lumber Beams Made of Beech. *BioResources* **2017**, *12*, 6122–6133. [CrossRef]
5. Nyström, J. Automatic measurement of fiber orientation in softwoods by using the tracheid effect. *Comput. Electron. Agric.* **2003**, *41*, 91–99. [CrossRef]
6. Simonaho, S.-P.; Tolonen, Y.; Rouvinen, J.; Silvennoinen, R. Laser light scattering from wood samples soaked in water or in benzyl benzoate. *Opt. Int. J. Light Electron Opt.* **2003**, *114*, 445–448. [CrossRef]
7. Purba, C.Y.C.; Viguier, J.; Denaud, L.; Marcon, B. Contactless moisture content measurement on green veneer based on laser light scattering patterns. *Wood Sci. Technol.* **2020**, *54*, 891–906. [CrossRef]
8. Jehl, A.; Bleron, L.; Meriaudeau, F.; Collet, R. Contribution of slope of grain information in lumber strength grading. In Proceedings of the 17th International Nondestructive Testing and Evaluation of Wood Symposium, Sopron, Hungary, 14–16 September 2011.
9. Jehl, A. Modélisation du Comportement Mécanique Des Bois de Structures Par Densitométrie X ET Imagerie Laser. Ph.D. Thesis, ENSAM, Cluny, France, 2012.
10. Viguier, J.; Bourreau, D.; Bocquet, J.-F.; Pot, G.; Bléron, L.; Lanvin, J.-D. Modelling mechanical properties of spruce and Douglas fir timber by means of X-ray and grain angle measurements for strength grading purpose. *Eur. J. Wood Wood Prod.* **2017**, *75*, 527–541. [CrossRef]
11. Olsson, A.; Oscarsson, J.; Serrano, E.; Källsner, B.; Johansson, M.; Enquist, B. Prediction of timber bending strength and in-member cross-sectional stiffness variation on the basis of local wood fibre orientation. *Eur. J. Wood Wood Prod.* **2013**, *71*, 319–333. [CrossRef]
12. Hu, M.; Olsson, A.; Johansson, M.; Oscarsson, J. Modelling local bending stiffness based on fibre orientation in sawn timber. *Eur. J. Wood Prod.* **2018**, *76*, 1605–1621. [CrossRef]
13. Hu, M.; Johansson, M.; Olsson, A.; Oscarsson, J.; Enquist, B. Local variation of modulus of elasticity in timber determined on the basis of non-contact deformation measurement and scanned fibre orientation. *Eur. J. Wood Wood Prod.* **2015**, *73*, 17–27. [CrossRef]
14. Viguier, J.; Bourgeay, C.; Rohumaa, A.; Pot, G.; Denaud, L. An innovative method based on grain angle measurement to sort veneer and predict mechanical properties of beech laminated veneer lumber. *Constr. Build. Mater.* **2018**, *181*, 146–155. [CrossRef]
15. Frayssinhes, R.; Girardon, S.; Denaud, L.; Collet, R. Modeling the Influence of Knots on Douglas-Fir Veneer Fiber Orientation. *Fibers* **2020**, *8*, 54. [CrossRef]
16. Lutz, J.F. Heating veneer bolts to improve quality of Douglas-fir plywood. Report No. 2182. In Proceedings of the Section Meeting of the Forest Products Research Society, Bellingham, WA, USA, 8–9 February 1960.

17. Stefanowski, S.; Frayssinhes, R.; Grzegorz, P.; Denaud, L. Study on the in-process measurements of the surface roughness of Douglas fir green veneers with the use of laser profilometer. *Eur. J. Wood Prod.* **2020**, 555–564. [CrossRef]
18. Mothe, F.; Marchal, R.; Tatischeff, W.T. Heart dryness of Douglas fir and ability to rotary cutting: Research of alternative boiling processes. 1. Moisture content distribution inside green wood and water impregnation with an autoclave. *Ann. For. Sci.* **2000**, *57*, 219–228. Available online: https://agris.fao.org/agris-search/search.do?recordID=FR2000003808 (accessed on 29 October 2020). [CrossRef]
19. NF-EN-322. Panneaux à Base de Bois-Determination de L'humidité. 1 June 1993. Available online: https://cobaz-afnor-org.rp1 .ensam.eu/notice/norme/nf-en-322/FA020887?rechercheID=1757547&searchIndex=1&activeTab=all#id_lang_1_descripteur (accessed on 8 June 2021).
20. OpenCV. Available online: https://opencv.org/ (accessed on 11 June 2021).
21. Tropix 7. Les Principales Caractéristiques Technologiques de 245 Essences Forestières Tropicales-Douglas. 2012. Available online: https://tropix.cirad.fr/FichiersComplementaires/FR/Temperees/DOUGLAS.pdf (accessed on 29 October 2020).
22. NF EN 384+A1. Structural Timber-Determination of Characteristic Values of Mechanical Properties and Density. 21 November 2018. Available online: https://sagaweb-afnor-org.rp1.ensam.eu/fr-FR/sw/consultation/notice/1539784?recordfromsearch= True (accessed on 29 October 2020).
23. Pollet, C.; Henin, J.-M.; Hébert, J.; Jourez, B. Effect of growth rate on the physical and mechanical properties of Douglas-fir in western Europe. *Can. J. For. Res.* **2017**, *47*, 1056–1065. [CrossRef]
24. Hankinson, R.L. Investigation of crushing strength of spruce at varying angles of grain. *Air Serv. Inf. Circ.* **1921**, *3*, 16.
25. Forest Products Laboratory and United States Departement of Agriculture, Forest Service. *Wood Handbook, Wood as an Engineering Material*; General Technical Report FPL-GTR-190; Forest Products Laboratory and United States Departement of Agriculture: Madison, WI, USA, 2010.
26. ImageJ. ImageJ Wiki. Available online: https://imagej.github.io/software/imagej/index (accessed on 23 June 2021).
27. Olsson, A.; Pot, G.; Viguier, J.; Faydi, Y.; Oscarsson, J. Performance of strength grading methods based on fibre orientation and axial resonance frequency applied to Norway spruce (*Picea abies* L.), Douglas fir (*Pseudotsuga menziesii* (Mirb.) Franco) and European oak (*Quercus petraea* (Matt.) Liebl./*Quercus robur* L.). *Ann. For. Sci.* **2018**, *75*, 102. [CrossRef]
28. Duriot, R.; Pot, G.; Girardon, S.; Denaud, L. New perspectives for LVL manufacturing wood of heterogeneous quality—Part. 2: Modeling and manufacturing of bending-optimized beams. to be defined.
29. NF EN 338. Structural Timber-Strength Classes. 1 July 2016. Available online: https://sagaweb-afnor-org.rp1.ensam.eu/fr-FR/ sw/consultation/notice/1413267?recordfromsearch=True (accessed on 29 October 2020).

Article

New Perspectives for LVL Manufacturing from Wood of Heterogeneous Quality—Part 2: Modeling and Manufacturing of Variable Stiffness Beams

Robin Duriot, Guillaume Pot, Stéphane Girardon and Louis Denaud

Arts et Metiers Institute of Technology, LaBoMaP, UBFC, HESAM Université, F-71250 Cluny, France; guillaume.pot@ensam.eu (G.P.); stephane.girardon@ensam.eu (S.G.); louis.denaud@ensam.eu (L.D.)
* Correspondence: robin.duriot@ensam.eu

Abstract: This paper presents a new strategy in the use of wood of heterogeneous quality for composing LVL products. The idea is to consider veneers representative of the resource variability and retain local stiffness information to control panel manufacturing fully. The placement of veneers is also no longer random as in the first part of this group of papers but optimized for the quality of veneers according to the requirement of bending stresses along the beam. In a four-point bending test arrangement, this means the high-quality veneer is concentrated in the center of the beam in the area between the loading points where the bending moments are the most important, and the low quality is located at the extremities. This initiates the creation of variable stiffness beams. This is driven by an algorithm developed and tested on representative veneer samples from the resource. Four LVL panels were manufactured by positioning the veneers in the same positions as in an analytical calculation model, which allowed the calculation of beam mechanical properties in four-point bending. The proposed optimization of LVL manufacturing from variable quality veneers should help for more efficient usage of forest resources. This optimization strategy showed notable gains for modeled and experimental mechanical properties, whether in terms of stiffness or strength. The analytical calculation of the local modulus of elasticity from modelized beams was satisfactory compared to the tests of the manufactured beams test results, allowing the reliability of the model for this property to be confirmed.

Keywords: laminated veneer lumber; Douglas-fir; heartwood; sapwood; bending optimization; veneer quality; four-point bending; local fiber orientation

Citation: Duriot, R.; Pot, G.; Girardon, S.; Denaud, L. New Perspectives for LVL Manufacturing from Wood of Heterogeneous Quality—Part 2: Modeling and Manufacturing of Variable Stiffness Beams. *Forests* **2021**, *12*, 1275. https://doi.org/10.3390/f12091275

Academic Editors: Ladislav Reinprecht and Ján Iždinský

Received: 15 July 2021
Accepted: 15 August 2021
Published: 17 September 2021

Publisher's Note: MDPI stays neutral with regard to jurisdictional claims in published maps and institutional affiliations.

1. Introduction

The laminated veneer lumber (LVL) is a lamellar composite material made of veneers from the rotary-peeling process. This material has several advantages. First, unlike sawn timber, it allows for a better distribution of defects within the material to avoid defects superimposition and thus homogenizes them in the product. Second, it can allow optimizing the placement of veneers according to their quality depending on the product's performance needs. This is a very important advantage, particularly on very heterogeneous woods in terms of density and number and size of defects, such as large Douglas-fir woods.

In the peeling industry, the sorting of veneers to differentiate their quality is usually performed according to appearance criteria (knot size, slots, resin pockets, etc.) based on EN 635-3 standard [1]. This classification is generally practiced by the intervention of cameras integrated on the panel-making line. In the first part of the series of articles [2], it was shown that local modeling of elastic properties from local fiber orientation and average density could establish quality sorting criteria, giving sorting results significantly different from the standard sorting method. Now, distributing this high-quality heterogeneity of veneers using their local mechanical properties to obtain the best possible performance of the material is a major issue. The scientific literature presents extensive work on the consideration of veneers of different quality in LVL. Until now, a number of works have been aimed at upgrading inexpensive or low-grade veneers with better quality veneers while maximizing the bending properties of the material. Different LVL stacking sequences mixing beech and poplar plies were also tested in edgewise bending by Burdulu et al. [3]. It is not surprising that the configuration with the greatest number of beech plies, the denser species, provided the best performance in terms of bending stiffness and strength. Chen et al. [4] also presented similar findings in a hybrid bamboo-poplar LVL. They also performed flatwise bending tests that highlighted the importance of veneer quality depending on their distance to neutral fiber in this configuration. Hybrid LVL was also tested in [5] with acacia mangium and rubberwood. This was exclusively done by varying the species according to the plies, showing improvements in flatwise bending stiffness and strength when more dense species were used. All these literature results dealt with sequencing different species or qualities of veneers in the thickness of LVL, but to the authors' knowledge, the optimization of the distribution of veneers of heterogeneous qualities along the product length has never been done before. It can be useful when LVL is used as slender beams loaded in edgewise bending.

This article belongs to a group of papers, whose first one is [2], which is dedicated to the use of heterogeneous species, such as large French Douglas-fir, in LVL manufacturing. In this first paper, the determination of local veneer properties by fiber orientation measurement integrated into the peeling line and veneer density enabled the establishment of a criterion to classify them according to their expected mechanical quality. Batches of veneers from the same classes were used to compose virtual panels with a random positioning of veneers. Panels were then virtually cut into beams, which were analyzed according to their equivalent modulus of elasticity to conclude the influence of sorting on the mechanical properties of the lamellar product. The aim was to propose an innovative grading method based on physical–mechanical properties and not just defects observed visually as it is performed for plywood industry (EN 635-3 standard), in order to manufacture reliable structural products.

However, the disadvantage of this random positioning of veneers is if the mechanical properties are rather low on average, consequently the mechanical properties of the LVL material will be low too. The sorting of the veneers can solve this issue but could lead to poor yields. However, in a beam subjected to bending, the stress is not homogeneous; thus, homogenized beams are partly oversized. Therefore, beams could be actually produced with heterogeneous mechanical properties to optimize utilization of raw material: using rather low-quality veneers where stress is low and high-quality veneers where stress is high. Mixing veneer qualities in a LVL beam is a challenge in the development of structural materials, as it means in this present context to consider the variability of the Douglas-fir resource to create beams with the highest possible mechanical properties.

The objectives of the present paper are multiple. First, a new strategy in the use of wood of heterogeneous quality for composing LVL products will be presented, enabling the quality of veneers to be optimized according to the requirement of bending stresses along the beam. This optimization is driven by a declination of a random veneer positioning panel manufacturing algorithm presented in [2]. Second, optimized panels were manufactured from a sample of veneers representative of the considered resource and were tested experimentally according to applicable standards [6,7]. The aim was both to draw conclusions on the model accuracy and to be able to estimate the performance of optimized beams realistically. In addition, to estimate the gains and losses in the mechanical properties of the optimized beams. Random beams from the same veneers were made numerically, simulating manufacturing without the view to optimizing panels. Finally, comparisons with beams from the different sorting methods from [2] were made so that the effects of grading could be estimated against the effects of optimized veneer placement.

2. Materials and Methods

First, this part explains how the veneers' longitudinal modulus of elasticity was calculated in relation to Part 1 [2], taking into account possible variations in the relationship between modulus of elasticity and density. Then, the algorithm developed to optimize the longitudinal position of the veneers within the laminate according to their quality is presented. Finally, the experimental manufacturing of optimized panels is described, as well as the equations involved in the calculation of mechanical properties, determined experimentally and analytically.

Table 1 contains the nomenclature listing the main symbols used.

Table 1. Nomenclature.

List of Main Symbols			
$\overline{\rho_{veneer}}$	Averaged veneer density at 12% moisture content (kg·m^{-3});	$\overline{E_{veneer}}$	Averaged local modulus of elasticity on all veneer surface (MPa);
$\theta(x,y)$	Local fiber orientation angle (°);	$\overline{E_x}(x)$	Stiffness profile: averaged local modulus of elasticity along the length of veneer (MPa);
$\overline{E_{beam}}(x,y)$	Average of $E_{ply,n}(x,y)$ of the 15 constitutive plies (MPa);	$\overline{E_{x,panel}}(X)$	Sum of veneer $\overline{E_x}(x)$ profiles across panel width, averaged by the sum of flags at each X-position (MPa);
$EI_{eff}(x)$	Effective bending stiffness profile along the length of the beam (N mm^2);	$\overline{E_{xmin}}$	Stiffness profile minimum value (MPa);
EI_{effmin}	Effective bending stiffness profile minimum value (N mm^2);	$E_x(x,y)$	Local modulus of elasticity of veneer (MPa);
$E_{m,l,an}$	Analytical local modulus of elasticity in bending (MPa);	$f_{m,exp}$	Experimental bending strength (MPa);
$E_{m,l,exp}$	Experimental local modulus of elasticity in bending (MPa);	$\overline{H_{veneer}}$	Averaged normalized local Hankinson value on all veneer surface (dimensionless quantity);
E_0	Theoretical longitudinal elastic of modulus for straight grain (MPa);	MC	Moisture content (%);

2.1. Pollet Equations

The Pollet Equation [8] was the basis of the LVL beam analytical modeling process. Determined from a Douglas-fir mature, clear wood specimen, it allows the calculation of a bending elastic module from density information. In [2], the longitudinal modulus of elasticity in clear wood of the veneers was calculated using the average veneer density at 12% of moisture content (MC) according to Equation (1).

$$E_0 = 36.605\,\overline{\rho_{veneer}} - 4242.4 \tag{1}$$

where:

- E_0 is the theoretical elastic of modulus for straight grain (MPa) at 12% MC;
- $\overline{\rho_{veneer}}$ is the veneer average density (kg m^{-3}) at 12% MC.

Then, the local fiber orientation regular grid was resized by applying a 12% MC tangential shrinkage coefficient [9]. Thus, the calculated mechanical properties of randomly composed beams were given at 12% MC to comply with convention.

In this paper, comparing the performance of manufactured LVL beams with modeled beams requires that veneer dimensions be consistent with their local fiber orientation regular grid. Indeed, material moisture content impacts these physico–mechanical properties (size, density, modulus of elasticity …) since the veneers were scanned in a green state, weighed after drying at 9% MC. Moreover, the beam moisture content was determined by double weighing small pieces of them during the destructive test [10], giving 7.5% MC for heartwood beams and 8% for sapwood ones. The Pollet equation (Equation (1)) requires 12% MC corrected density. Local longitudinal modulus of elasticity at 12% MC $E_x(x,y)$ was expressed from local fiber orientation regular grid resized in veneer width direction with a 7.5% (respectively 8%) MC tangential shrinkage coefficient for heartwood (respectively sapwood). Thus, the mechanical properties calculated from analytical tests were given at 12% MC. Concerning experimental moduli, they were corrected at 12% MC according to [11].

The experimental scatter used to calculate Equation (1) from this linear regression was very sparse, which raises the question of the accuracy of the formula. Thus, to consider the entire domain of potential modulus/density relationships, some declinations of it were calculated. The slope coefficients were all taken equal to the 36.605 m^2s^{-2} value of Equation (1), and several Y-intercept values were recalculated based on the percentage of points below each linear equation, as showed in Figure 1.

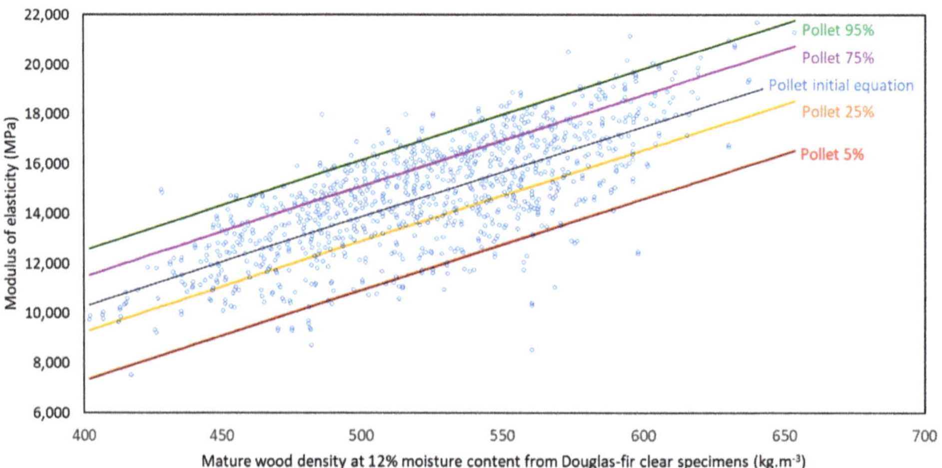

Figure 1. Linear regressions from the relationship between density and longitudinal modulus of elasticity for mature Douglas-fir clear specimens from Pollet [8].

It means that, in the example of the 75% Pollet equation, the Y-intercept value was calculated in order that 75% of points were under the line. Table 2 presents the coefficients of these equations. Thus, for example, the 75% Pollet equation implies +1052.6 MPa to add to the value of the longitudinal elastic modulus of clear wood.

Table 2. Pollet linear regression variants.

Percentage of Points under Line	Slope Coefficient	Y-Intercept	Stiffness Gain/Loss (MPa)
95%		−2140.1	+2102.3
75%		−3189.8	+1052.6
Initial equation	36.605	−4242.4	0
25%		−5408.8	−1166.4
5%		−7368.8	−3126.4

The results from destructive tests presented afterward will be put into perspective with these equations to consider the sensitivity of Equation (1) in the mechanical properties prediction.

2.2. Favorable, Unfavorable, and Random Optimization Positioning

The presence of knots between loading points of the neutral axis creates zones of weakness that initiate failure. In addition, low density leads to low static bending strength [8]. The main objective was, therefore, to minimize the presence of zones of mechanical losses due to high knottiness and/or low density between loading points in a four-point bending configuration. The presence of knots in the beam area under tensile solicitation is critical, but this optimization was done in an undifferentiated manner between areas of compression and tensile in order to not limit by the direction of the beam in its use. A model for optimizing the distribution of veneers was developed to improve the mechanical properties of beams according to the bending stress imposed. The principle of optimizing the distribution proceeded by placing the veneers from the center of the panel towards the extremities and selecting the best performing veneer available on a defined criterion. The expected mechanical consequences were, therefore, better strength and higher local elastic moduli due to lower local grain angles in the beam area between the loading points.

Figure 2 is a flowchart describing the veneer-positioning algorithm, whose similarities can be observed with the functionality of the random placement algorithm presented in [2]. Boxes representing new functions are framed in red.

As in Part 1 [2], $(\vec{X}, \vec{Y})$ was the coordinate system of the panel. Each of these positions was initially considered as available. All vacant positions were first classified in ascending order according to their eccentricity distance (noted $X_{i,eccentricity,theo}$ in [2]). The first vacant position to be assigned was the one with the lowest absolute value of the eccentricity distance. Then, the highest-criterion veneer of the highest class was selected and allocated to this position. At this stage, the term criterion needed to be defined.

Here, the hierarchy between veneers of the same class was made according $\overline{E_{veneer}}$ value, defined in [2]. Each of the veneers was then picked to fill each of the vacant positions progressively. A feature of nonsuperposition of areas of low elastic modulus between plies was developed and added to avoid creating local mechanical weaknesses in the width of the material. The principle is explained in Figure 3. Each time a veneer was placed in a vacant position, a test was performed to choose between $0°$ or $180°$ to minimize defect superimposing. A sum of all stiffness profiles across the 15 plies between the X abscissa of the edges of the considered position (X_{min} and X_{max}) was performed. A flag system allowed this localized profile to be divided by the number of veneers allocated at this stage on the 15 plies on each X-axis coordinate between X_{min} and the X_{max}: its designation was $\overline{E_{x,panel}}(X)$. Then, the highest value found between the 2 cases dictated the positioning angle.

As in Part 1 [2] random algorithm, each time a veneer was allocated to a position, the theoretical abscissa of the vacant positions was recalculated from its real length. To simplify the algorithm, a veneer attributed to an end of ply was cut in half as required, and the second part completed the opposite ends. Again, for each one, the condition of no superimposing of defects was applied. Once all veneers had been assigned, and therefore all plies of the material panel were filled, numerical sawing of the beams in the panel was done according to the input setting dimension.

In order to put the mechanical results of the favorably optimized beams into perspective, the working principle of the algorithm previously described was reversed in order to also be able to create the opposite of favorably optimized beams. The principle of assigning positions in ascending order of eccentricity distance from the center of the panel was retained, but the choice by veneer performance was reversed. Veneers were selected and placed from worst to best quality according to prioritization criteria, the $\overline{E_{veneer}}$ value.

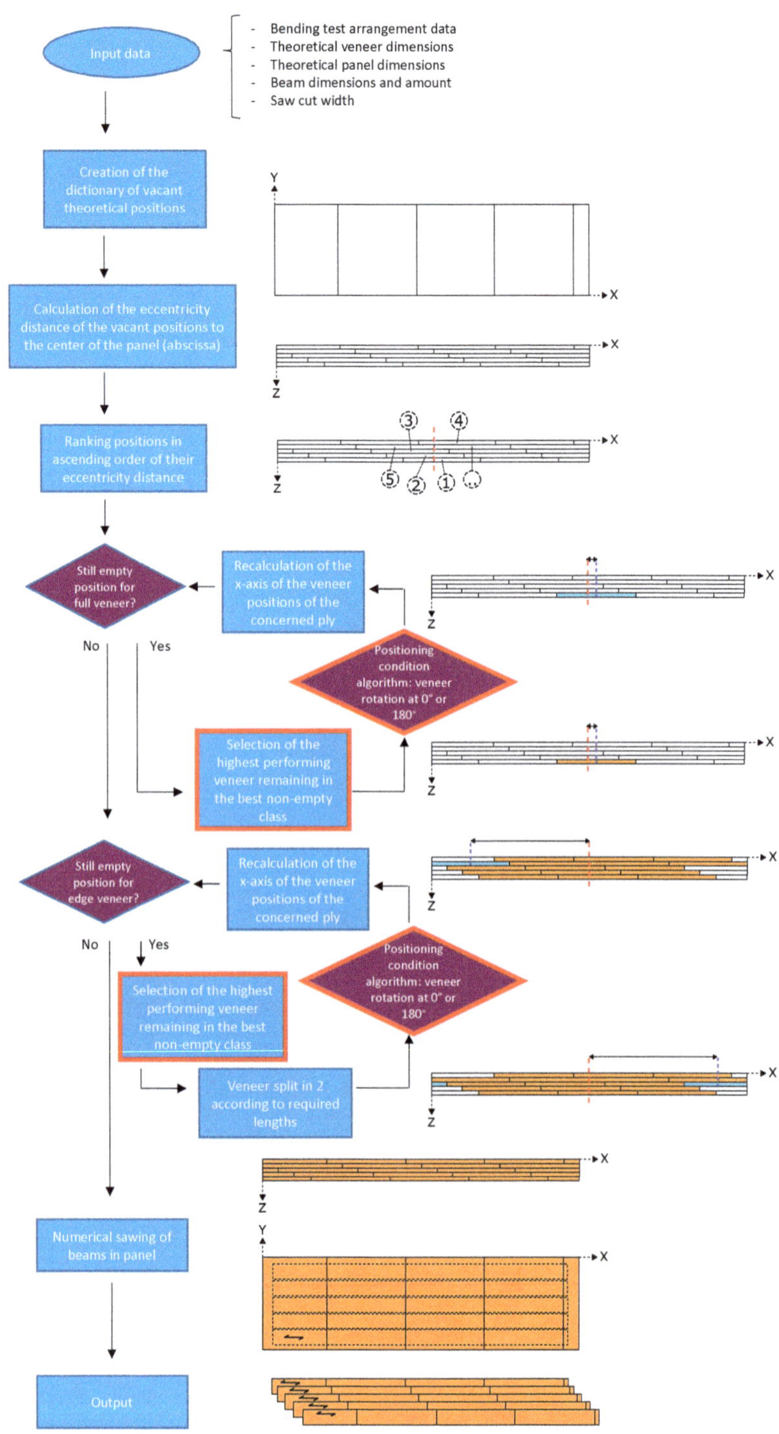

Figure 2. Favorable placement veneer algorithm flowchart.

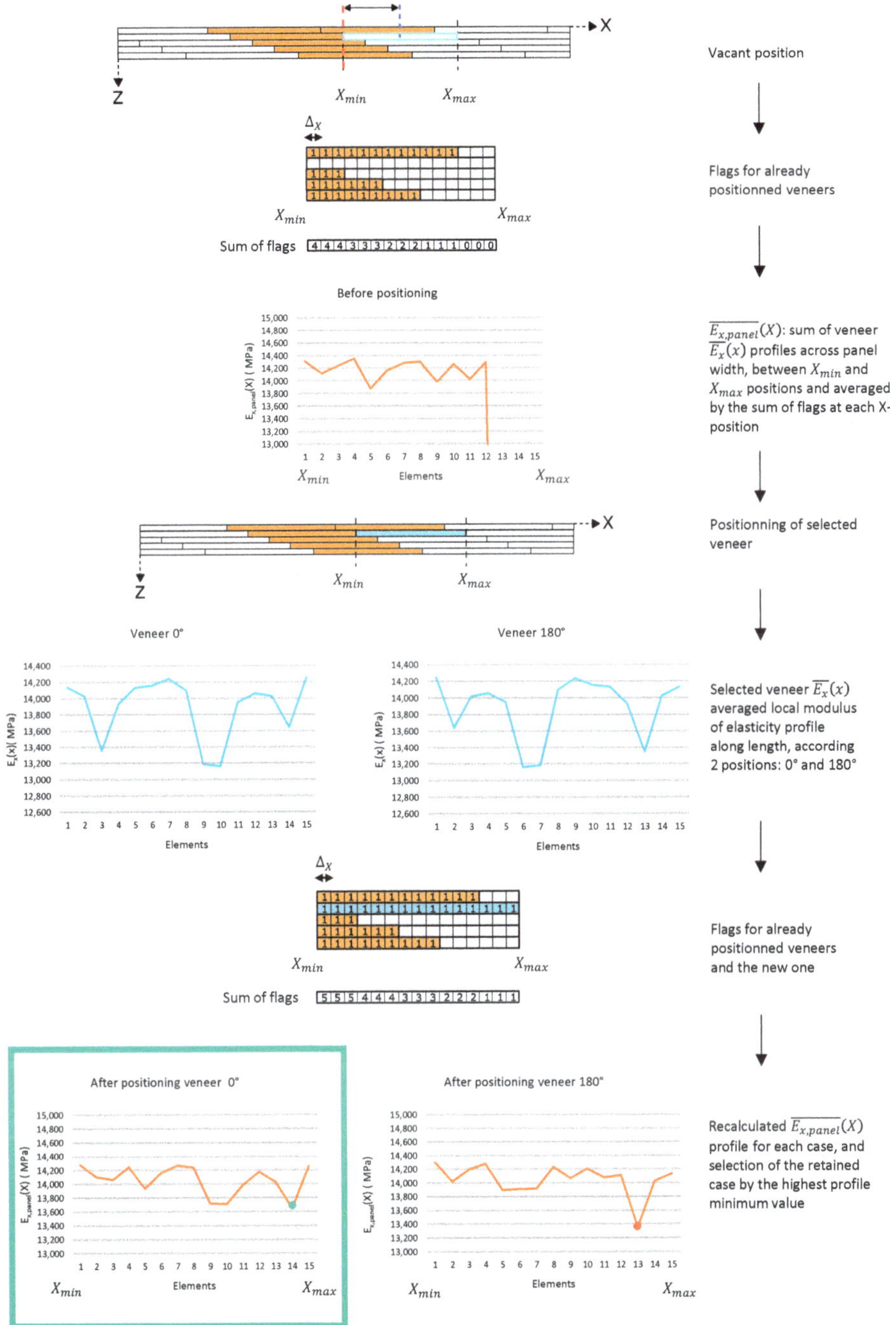

Figure 3. Nonsuperposition principle.

The Part 1 [2] numerical panel manufacturing algorithm by random veneer positioning was used too in order to evaluate the performance of optimized beams by generating panels and random beams. However, it was not possible to generate all the combinations of possible panels. Indeed, in this study, a panel was composed of 60 full veneers, divided into 72 positions considering veneers at extremities cut in half. Each veneer had two possible positions: $0°$ and $180°$. In other words, there were $60! \times 2^{72} = 3.93 \times 10^{100}$ possible combinations. Obviously, not all these configurations could be computer-generated in a reasonable time as it takes one minute per panel. For each batch of veneers, which were used to compose a panel, a sample of 150 combinations was generated per panel to allow some conclusions. A statistical test presented later will show that this hypothesis is relevant.

2.3. Panel Manufacturing and Samples Preparation

2.3.1. Composition of Representative Resource Batches

To allow for comparison of the mechanical results of the beams from heartwood and sapwood panels, a batch sampling of the 60 necessary veneers was conducted based on their individual $\overline{E_{veneer}}$ criterion. The objective was to obtain 2 batches of heartwood and 2 batches of sapwood veneers as similar as possible between them according to this criterion.

For this study, 4 optimized panels were composed:

- Heartwood batch n°1 veneers were used to manufacture a favorable veneer placement optimization without the "defect nonsuperposition" option: dense clear wood was concentrated in the middle of the panel;
- Heartwood batch n°2 veneers were used to manufacture a favorable veneer placement optimization, with the "defect nonsuperposition" option: veneer positioning was driven by their quality and the homogenization of defects by the algorithm portion shown in Figure 3;
- Sapwood batch n°1 veneers were used to manufacture a favorable veneer placement optimization without the "defect nonsuperposition" option: all available good quality wood veneers were concentrated in the center of the panel;
- Sapwood batch n°2 veneers were used to manufacture an unfavorable veneer placement optimization: all available low-quality wood veneers were concentrated in the center of the panel, with the "defect superposition" option: in the final step of the algorithm in Figure 3, the lowest value of $\overline{E_{x,panel}(X)}$ was chosen to maximize the accumulation of areas of low mechanical property.

Once optimized (favorably or unfavorably), veneer positioning algorithms were used to build the panels, and a virtual cut-out of each beam in their actual dimensions and respecting placement in the real panel was performed.

In order to compare the sapwood and heartwood beams, respectively, the sampling of the veneers had to meet two requirements: the batches, representative of the initial resource, must have an overall equivalent population distribution by wood type. In practice, it is important to remember that this is impossible since each veneer is unique. To compose the veneer samples, the averaged local modulus of elasticity on all veneer surfaces $\overline{E_{veneer}}$ was used as a quality criterion.

However, this criterion alone was not sufficient to attest to the proper equivalency of batches. The two parameters involved in the determination of the local modulus of elasticity of veneer, which were fiber orientation angle and density (as specified in [2]), were also observed. Table 3 shows these parameters for the complete veneer resource and the 2 sampled batches for each type of wood.

Table 3. Batches equivalence according to considered criteria (12% MC).

| | | Heartwood | | | Sapwood | | |
| | | Mean Value (COV %) | | | Mean Value (COV %) | | |
	Unit	Sample	Batch n°1	Batch n°2	Sample	Batch n°1	Batch n°2
Amount of veneers	-	163	60	60	123	60	60
$\overline{\rho_{veneer}}$	kg.m^{-3} (%)	513 (5.6) b	506 (5.3) b	511 (5.1) b	559 (5.0) a	559 (4.6) a	559 (5.6) a
$\overline{H_{veneer}}$	d.q. (%)	0.92 (4.5) a	0.94 (3.8) a	0.92 (4.2) a	0.93 (4.8) a	0.93 (4.8) a	0.94 (4.6) a
$\overline{E_{veneer}}$	MPa (%)	13,349 (7.4) b	13,342 (6.9) b	13,284 (7.0) b	15,112 (7.4) a	15,082 (7.3) a	15,161 (7.5) a

Values followed by a different letter within a row are statistically different at a 5% *p*-value (ANOVA and Tukey HSD test).

According to the Tukey HSD test, when comparing sapwood and heartwood veneers independently, there were no significant differences between batches and no statistically significant difference between batches compared to samples, either on the overall individual veneer density, the average standardized Hankinson value, or on the average veneer modulus of elasticity. This, therefore, ensured that the batches used to create optimized panels were globally equivalent according to these criteria and thus allowed for the comparison of properties between them.

2.3.2. Panel Manufacturing

Four panels were produced according to veneer positioning as identical as possible to the models. A measuring tape was used to check the positioning of the veneers in the $\overrightarrow{X}$ length of the panel, according to the $(\overrightarrow{X}, \overrightarrow{Y})$ coordinate system. A stop along the length of the panels enabled positioning the veneers at Y=0. The accuracy of this positioning after gluing was estimated to +/−2.5 mm. Because of this and because the veneer surfaces were not always rectangular, areas of mounted joints (superposition of two veneers that locally cause a double veneer thickness) and cages (local gap between veneers' borders that cause a hole in the panel) were observed, not exceeding 4 mm in length. The gluing process was made with a vacuum press (woodtec Fankhauser GmBh) [12], shown in Figure 4a The glue used was a polyurethane glue (PU) with a spread rate of 200 g·m^{-2} (Figure 4b). Panels were manufactured two by two (Figure 4c). The glued veneers had a nominal thickness of 3 mm. Their nominal length was between 700 or 750 mm, according to the length of each log, and their cutting width was 850 mm or 1000 mm, adjusted according to the external aspect of the log (visible knottiness, radial cracks). They were pressed together by a diaphragm vacuum press combined with a vacuum pump producing a pressure of 0.08 N mm^{-2} for 8 h (Figure 4d). After gluing and pressing, the LVL were stacked and stabilized for one week before cutting.

The bending beams were cut from the panels with a panel saw. The nominal dimensions of each beam were 2710 mm × 145 mm × 45 mm. Five beams were cut from each panel. The positioning of beams in the panels was measured to match with the coordinate entered in the algorithm. Because the edges of the panels were not perfectly conspicuous due to the imprecision of veneer placement, an accuracy between 5 mm and 10 mm was estimated in the (X,Y) coordinate positioning of beams in the panel.

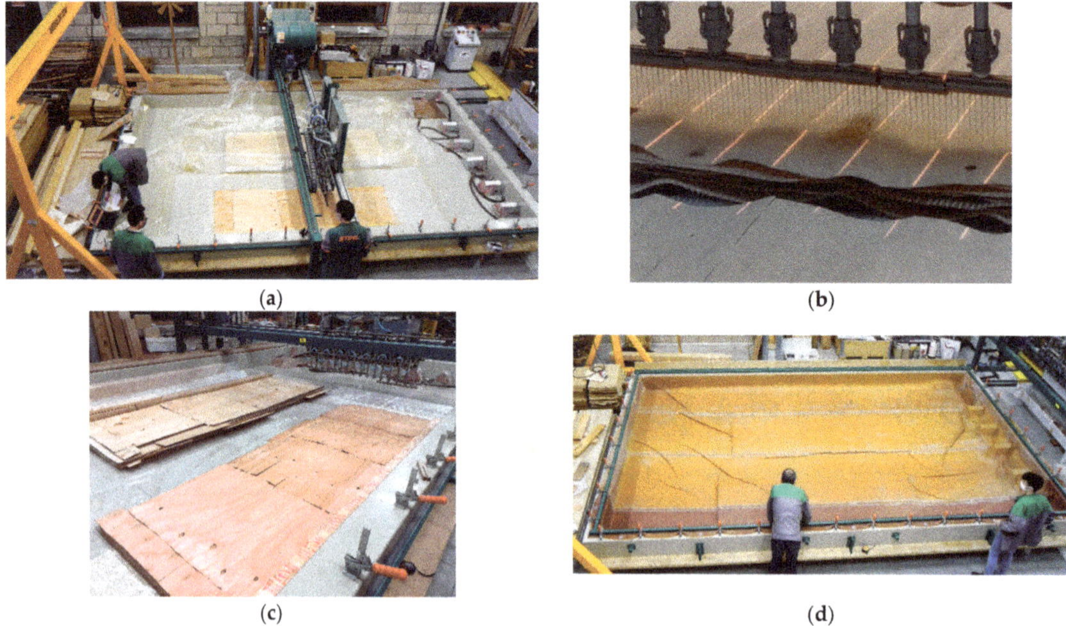

Figure 4. Panel manufacturing: (**a**) general view, (**b**) glue nozzles, (**c**) veneer positioning of plies, (**d**) pressing.

2.4. Bending Test Mechanical Properties Assessment

All the specimens were tested in the edgewise direction only. This choice was made to guarantee a sufficient number of bending tests in this direction which is the one used for LVL beams as a slender flexural product. In addition, this test configuration presents two advantages for a fair comparison of the material properties: all the plies were subjected substantially to the same mechanical loading, and the glue joints between plies were globally less subjected to stresses than in flatwise bending. The four-point bending tests were made with a dedicated testing machine composed of an electric actuator equipped with a 100 kN load cell. The upper and lower supports were made of 4.8-cm wide metal plates, fixed on a pivot, allowing the rotation of the beam supports coated with PTFE material.

To get mechanical properties, a four-point bending test was performed on every specimen, modelized and manufactured, based on the EN 14374 [6] and EN 408+A1 [7] standards. A distance equal to 2610 mm, as 18 times the specimen height between the lower supports of 145 mm, was set, as shown in Figure 5. The distance between the loading head and the nearest support (a) was set to 870 mm, which was 6 times the height, in order to prevent possible shear failures. The distance between upper support was also 6 times the height to conform to standards. The destructive test local deflection was measured with a LVDT sensor in 5 times height length portion, between 6.5 h and 11.5 h length from the first lower support.

The 4-point bending analytical calculation model was adapted from the one developed in [13] in order to switch from flatwise bending to edgewise bending.

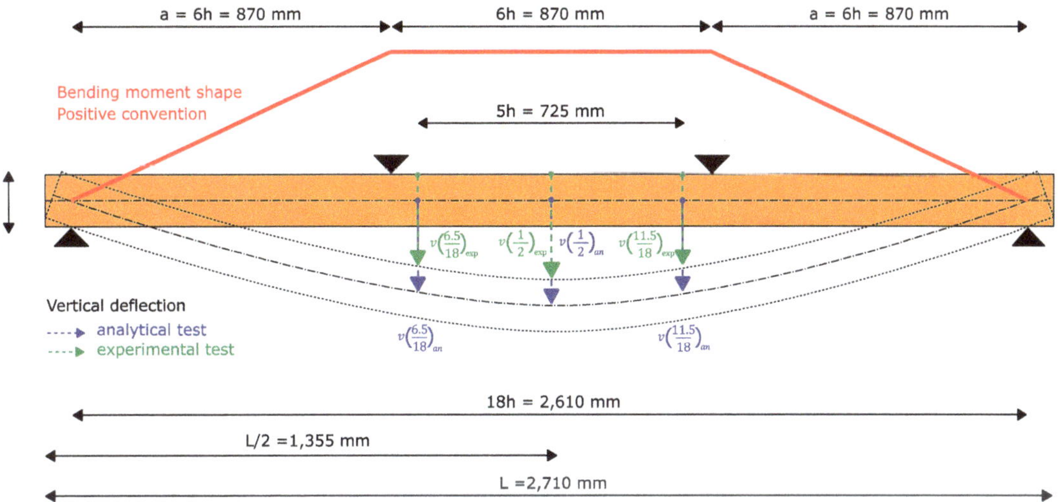

Figure 5. Four-point bending test arrangement.

From the $EI_{eff}(x)$ beam equivalent stiffness curve (defined in [2]), the deflection at mid-span $v\left(\frac{1}{2}\right)_{an}$ of the modeled beams for a four-point bending test of the modeled panels was calculated using the Müller–Breslau's principle (see Equation (2)).

$$v\left(\frac{1}{2}\right)_{an} = \sum_{i=1}^{n_x} \frac{M_{f,i}(x) M_{v,i}(x)}{EI_{eff,i}(x)} \Delta x \tag{2}$$

where:

- M_f is the bending moment during a four-point bending test induced by the application of unitary load at x position of loading points (MPa);
- M_v is the bending moment induced by a unitary load at a given x position (mid-span in this case) (MPa);
- EI_{eff} is the effective bending stiffness calculated previously, which is dependent on the local modulus of elasticity (N mm^2);
- n_x is the number of discrete elements along $\vec{x}$ direction;
- Δ_x (1 mm) corresponds to the resolution of the interpolated images along the $\vec{x}$ direction.

To be consistent with the experimental measurements, the deflection at the 2 support points used to measure local deflection was also calculated according to Equation (2). The mean value of the 2 deflections $v\left(\frac{6.5}{18}\right)_{an}$ and $v\left(\frac{11.5}{18}\right)_{an}$ were subtracted from the mid-span deflection to get the local relative deflection, as shown in Equation (3).

$$v_{l,an} = v\left(\frac{1}{2}\right)_{an} - \frac{v\left(\frac{6.5}{18}\right)_{an} + v\left(\frac{11.5}{18}\right)_{an}}{2} \tag{3}$$

The experimental and analytical local modulus of elasticity was calculated according to the beam theory in 4-point bending using Equation (4) of Table 4 given in the NF-EN-408+A1 standard [7].

Table 4. Analytical and experimental local modulus of elasticity equation.

Local Modulus of Elasticity		
$$E_{m,l} = \dfrac{al_1{}^2}{16I_{Gz}\frac{\Delta w}{\Delta F}}$$		(4)
$E_{m,l,exp}$	$E_{m,l,an}$	

where:

- l_1 is the gauge length for the determination of modulus of elasticity (mm)
- $\Delta F = F_2 - F_1$ with $F_1 = 0.1\,F_{max}$ and $F_1 = 0.4\,F_{max}$ is the increment of load on the linear regression with a correlation coefficient of 0.99 or better (N),
- $\Delta w = w_2 - w_1$ is the increment of deformation in millimeters corresponding to ΔF, measured by the local deflection LVDT sensor (mm).

- ΔF is the unitary load which induced the previous bending momentum M_f (N),
- $\Delta w = v_{l,an}$ the mid-span local deflection term is the one calculated by Equation (2).

Contrary to the requirements of the calculation bounds of the modulus for the destructive test, the assumption of the pure elastic behavior of the material considered in the analytical model allowed the local modulus to be calculated on a unitary loading increment.

During the experimental testing of the beams, the bending strength of each beam was measured. The bending strength was calculated according to EN 408+A1 [7] and Equation (5).

It is usual to assume that low stiffness in timber is associated with low strength. Olsson et al. did this in [14] for Norway Spruce timber, also Viguier et al. in [15]. Thus, the lowest bending stiffness computed from the analytical model as defined in Part 1 [2] can be used as a criterion to correlate to the experimental bending strength. The criterion, symbolized E_{min}, which consisted of calculating the minimum value of $EI_{eff}(x)$ in the between the upper-support portion of beam length (6 times the height of the beam), and then redividing by the nominal moment of inertia to enable an easier reading of results (see Equation (6) of Table 5).

Table 5. Bending strength indicating properties.

Bending Strength			
$f_{m,exp} = \dfrac{a\,F_{max}\,h}{4\,I_{Gz}}$	(5)	$E_{min} = \dfrac{\min(EI_{eff}([x_{lus}:x_{rus}]))}{I_{Gz}}$	(6)

- F_{max} is the maximum bending effort given by the load sensor (Newton);
- a is the distance between a loading point and the nearest support (mm);
- h is the nominal beam height (mm);
- a is the distance between a loading point and the nearest support (mm);
- I_{Gz} is the moment of inertia for a rectangular cross-section beam (mm^4) defined as:

$$I_{Gz} = \frac{bh^3}{12}$$

- $EI_{eff}(x)$ is the effective bending stiffness profile along the length of the beam (N mm^2);
- x_{llp} is the left loading point x-position in the beam system of axis;
- x_{rlp} is the right loading point x-position in the beam system of axis.

It has already been used in literature as indicating property to bending strength [14]. For timbers, a moving average is usually calculated. In the case of the LVL, this value was considered already averaged because the $\overline{E_{beam}}(x,y)$ regular grid used for the $EI_{eff}(x)$ calculation was already an average over the 15 plies of the material.

3. Results

3.1. Effect of Beam Optimization on Modeled Mechanical Properties

First, observations can be made on the impact of the controlled distribution of veneers based on their quality within the different panels on their effective bending stiffness.

The colored $EI_{eff}(x)$ curves presented in Figure 6a are the bending stiffness profiles for each modeled and manufactured beam. They allow for a very explicit observation of the effects of the veneer positioning processes. Favorably optimization led to observing maximum bending stiffness values in the central part of the beams and a regularly decreasing bending stiffness at the ends of the beams. The opposite was also observable for unfavorably optimized beams. The dotted vertical lines represent the loading points of the four-point bending test arrangement. The black curves in Figure 6b,c,f,g represent the $EI_{eff}(x)$ curves of beams obtained from random positioning of veneers in the panel. The 150 cases per batch simulated represent 750 beams (5 beams cut for each panel). All displayed black curves in Figure 6b,c,f,g, and showed the possible "area of performance". This allowed showing how the optimized bending stiffness profiles provide the best possible stiffness to fit the applied bending moment, while the random arrangements had up and downs randomly all along the beams (shown by individual grey curve examples). Figure 6d,e,h,i are examples of colormaps of the local modulus of elasticity averaged over the 15 plies of LVL beam generated from different batches ($\overline{E_{beam}}(x,y)$, as defined in Part 1). Figure 6d,h,i shows a concentration of green between loading points, corresponding to higher local modulus than in the beam end areas. There was also a variation in the color range between the sapwood beams (Figure 6d,e) and heartwood ones (Figure 6h,i), explained by a density effect between the two types of wood.

For cases of favorably optimized beams, an $EI_{eff}(x)$ profile was observed, with higher local values at the center of the beams than at their extremities, which corresponded well to a positioning of the veneers with lower quality at the ends and those of higher quality at the center (Figure 6b,f,g). On the contrary, for unfavorably optimized beams, a concentration of local high $EI_{eff}(x)$ values at extremities and lower-quality veneers at the center was observed in Figure 6c.

Comparing heartwood and sapwood, a general gap between profiles was observable on the optimized profiles. This difference was mostly imputable to a difference in the veneers' density. To estimate this difference, an overall average of all the points in the profiles of all the sapwood and heartwood beams was performed. The average value was $1.54.10^{11}$ N mm^2 for heartwood and $1.74.10^{11}$ N mm^2 for sapwood (difference of 13.0%).

There was a difference between the positioning of veneers leading to the favorably optimized heartwood curves in Figure 6f,g. For the first one, there was a high local maximum between loading points (slightly above the "area of performance"), but also a low local minimum between loading points compared to the curves of the second batch. This was due to the placement algorithm, modified between the creation of the two batches to improve strength. In the first case, the requirement to place the higher-quality veneers in the center was preponderant, while in the second case, the effect of maximizing the plateau value between the loading points was sought.

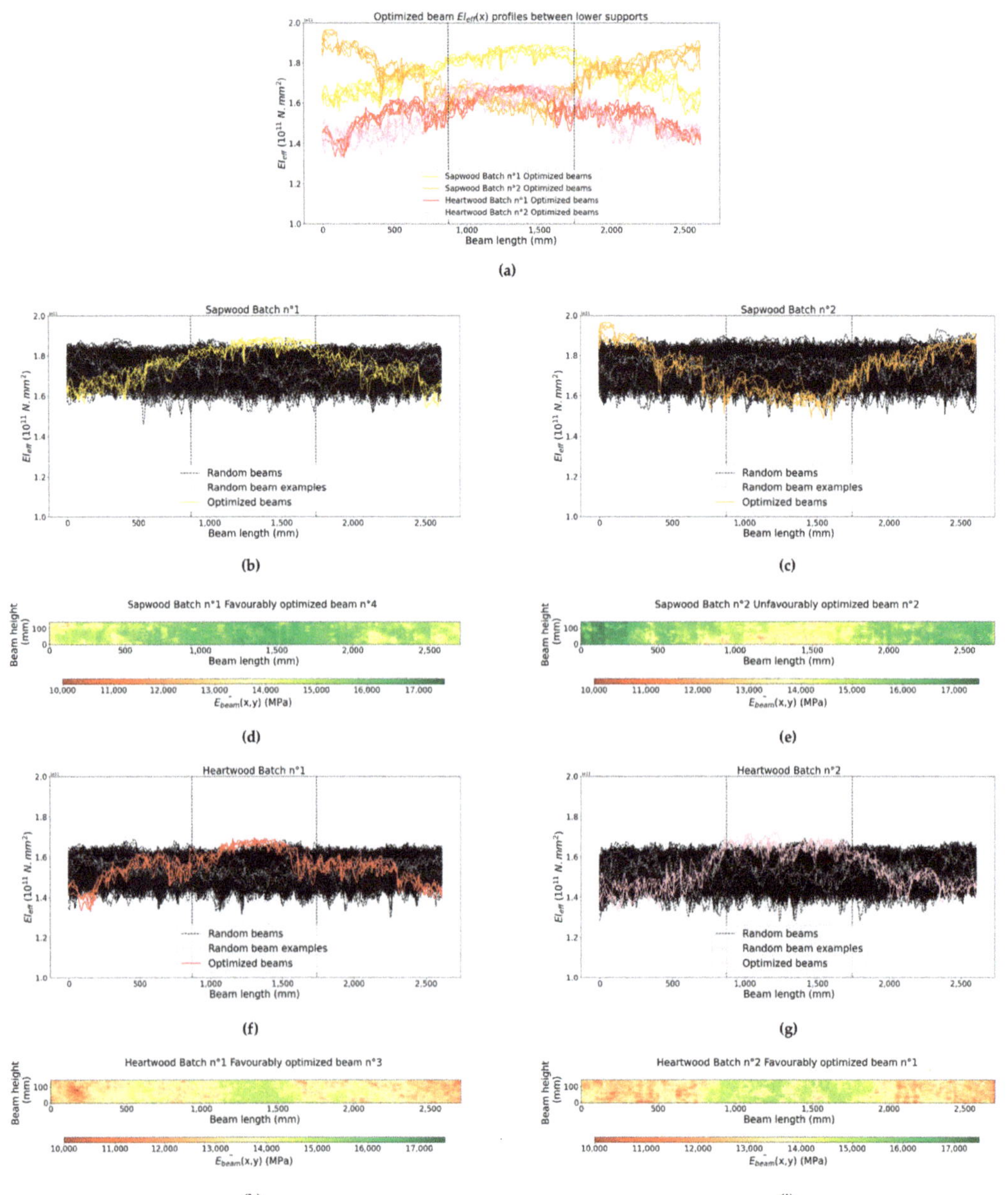

Figure 6. Beam EI_{eff} profiles: (**a**) optimized beams from the four batches, (**b**) batch n°1 sapwood beams among random beams, (**c**) batch n°2 sapwood beams among random beams, (**f**) batch n°1 heartwood beams among random beams, (**g**) batch n°2 heartwood beams among random beams. Beam $\overline{E_{beam}}(x,y)$ grid: (**d**) batch n°1 sapwood beam, (**e**) batch n°2 sapwood beam, (**h**) batch n°1 heartwood beam, (**i**) batch n°2 heartwood beam.

3.2. Modeling Result Analysis by Comparison with Destructive Results

To analyze the accuracy of the mechanical properties modeling of beams, a comparison was made between the results of destructive testing on the optimized beams and analytical tests on their numerical equivalent.

$E_{m,l,ex}$ experimental results were recalculated to 12% MC according to EN 384 standard, to be compared with $E_{m,l,an}$. These results are visible in Figure 7. Whisker plots enabled us to observe the domain of mechanical properties, whose edges represent the value resulting from the application of the 25% and 75% variants of the Pollet equation from Table 2.

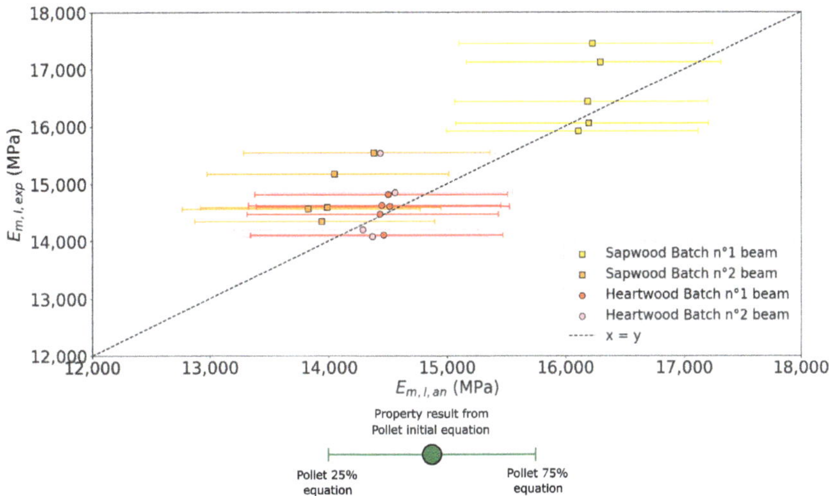

Figure 7. Scatter plot between analytical and experimental local modulus of elasticity.

The relative differences between modelized results from Pollet [8] mature Douglas wood equation (Equation (1)) according to experimental results were on average of −0.9% for heartwood, with a minimum deviation of −7.1% and a maximum deviation of +2.5%. For sapwood, the difference was, on average, −3.8%, with a minimum deviation of −7.5% and a maximum of +1.2%. Thus, despite the fact that this equation is applicable for mature wood, the analytical bending modeling was found to have a tendency to undervalue the local modulus of elasticity, especially for sapwood, and more clearly for unfavorable optimization. Moreover, the model exhibited difficulty perceiving the singularities of the different beams from the same panel. Hence, it tended to be relatively consistent in its predictions ($R^2 = 0.73$), but the conclusions cannot be drawn any further given the reduced number of specimens.

The experimental tests showed the optimization effect of the panels on the $E_{m,l,exp}$ and $f_{m,exp}$ results of the four batches (shown in Table 6).

Table 6. Experimental results.

			Heartwood Mean Value (COV %)			Sapwood Mean Value (COV %)			
			Configuration	$E_{m,l,exp}$ (MPa)	$f_{m,exp}$ (MPa)		Configuration	$E_{m,l,exp}$ (MPa)	$f_{m,exp}$ (MPa)
Representative veneer batches	Batch n°1	5 beams	Favorable optimization	14,527 (1.8)	50.96 (8.2)	5 beams	Favorable optimization	16,599 (4.0)	61.47 (6.8)
	Batch n°2	4 beams	Favorable optimization	14,665 (4.6)	49.30 (11.9)	5 beams	Unfavorable optimization	14,847 (3.3)	47.99 (11.3)

Indeed, the average relative deviation for $E_{m,l,exp}$ between sapwood favorably and unfavorably optimized beams was +11.8% (16,599 MPa vs. 14,847 MPa), while the relative deviation between the $E_{m,l,an}$ mean value was +15.5% (16,205 MPa vs. 14,036 MPa). The relative deviation in between number 2 and number 1 batches of heartwood-favorably optimized beams $E_{m,l,exp}$ mean value was +1.0% (14,665 MPa vs. 14,527 MPa) while the relative gap between the $E_{m,l,an}$ mean value was −0.4% (14,413 MPa vs. 14,472 MPa). The absolute values were also very close, highlighting the great calibration of analytical model parameters, as noticeable in Figure 7.

Despite unavoidable variations between beams, these average results provided some confidence in the representativeness of the modeling method, allowing for deeper exploitation of modeling results in the following section.

Regarding bending strength in sapwood batches, there was an average relative deviation of +28.1% between the favorably optimized and unfavorably optimized batches. This value is remarkable given that only the positioning of veneers according to their quality differs between the batches. Considering the small sample of beams tested, it was difficult to conclude the predictive aspect of the model by the E_{min} criterion on the strength of beams tested in four-point bending, as shown in Figure 8.

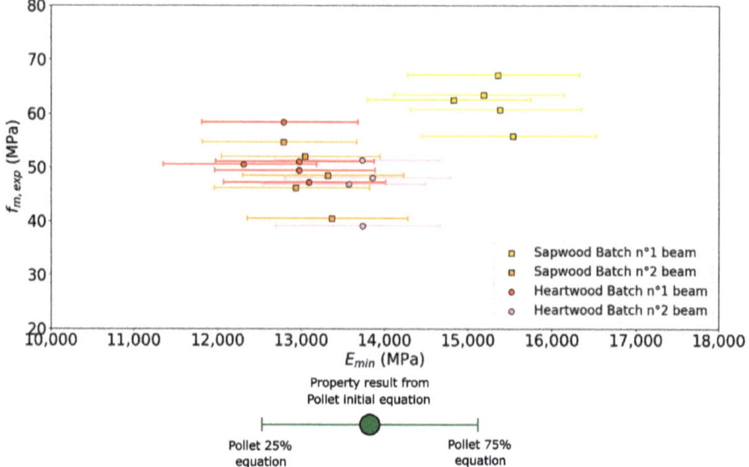

Figure 8. Scatter plot between E_{mini} between loading points from modelization and experimental bending strength.

However, the effect of favorable and unfavorable optimizations on the beams of the batches was noticeable, as for local modulus of elasticity. For batch n°2 sapwood unfavorably optimized beams, the $f_{m,exp}$ bending strength was between 40 MPa and 52 MPa, while for batch n°1 favorably optimized beams, it was between 55 MPa and 67 MPa. Concerning the E_{min} criterion, batch n°2 optimized beams averaged 13,091 MPa while batch number 1 optimized beams averaged 15,257 MPa (+16.5%). For heartwood, the two favorably optimized batches presented similar strength between 39 MPa and 59 MPa. It seems that, even if an increasing global trend was observed, the proposed criterion was not able to differentiate precisely the observed variation of bending strength for all favorable or unfavorable LVL composition.

3.3. Comparison with Sorted and Unsorted Random Beams

As a point of comparison to evaluate the performance of optimized beams, random panels created using the algorithm developed in [2] were generated from the same veneer batches. Figures 9 and 10 show respectively $E_{m,l,an}$ and E_{min} of beams with optimized veneer placements compared to random beams made from the same veneers. Violin plots allows observing the distribution law of random values. Dots representing optimized beams enabled us to observe the optimization of the beams in relation to the random sample.

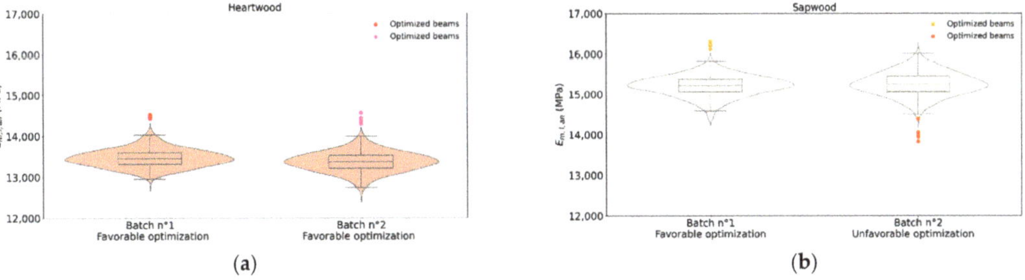

Figure 9. Local elastic modulus in bending results according to cases: (**a**) heartwood batches, (**b**) sapwood batches.

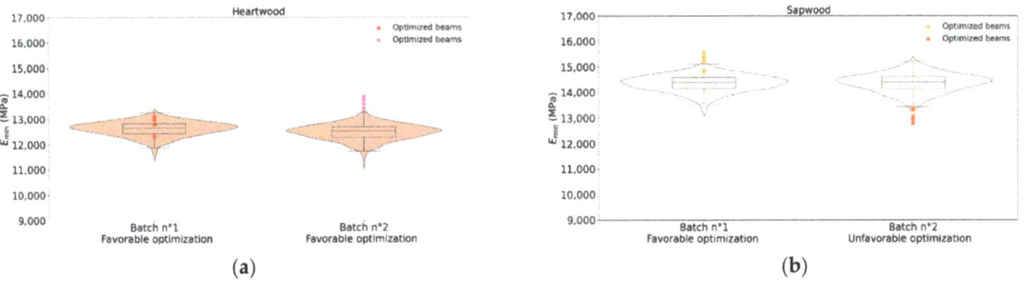

Figure 10. E_{mini} between loading points according to cases: (**a**) heartwood batches, (**b**) sapwood batches.

For each sample of 150 random beam results, a Shapiro–Wilk test was performed. For $E_{m,l,an}$ and E_{min} random beams results, a Gaussian distribution was observed (p-value < 5%). This enables the use of 150 random beam sample trends to be relevant to conclusions in general terms. For E_{min} results, the distribution was crushed over the higher values. This was due to the nature of the parameters. The local modulus elasticity was calculated for a portion of beam (5 h length). This gave a kind of averaged value. The E_{min} value was by nature a minimum value for substantially the same portion of beam (6 h length).

Table 7 summarizes the main results of the study, including Part 1 results. It compares the results of beams from optimized panels with random panels, from the same batches, and random beams of [2], made of veneers of different sorting classes. The E_{min} values are presented with mean values, but also by their 5th percentile value, to continue the analogy with the strength property usually used in the construction design. In the case of the manufactured beams, since optimized batches were composed of only five beams maximum, the minimum of E_{min} values was used.

First, a comment on the results of the random heartwood and sapwood batches beam test results can be made. For heartwood, there was a relative deviation of −0.7% between $\overline{E_{m,l,an}}$ of batch number 2 random beams (13,362 MPa) and batch number 1 random beams (13,462 MPa). For $\overline{E_{min}}$, the relative deviation was on average −1.2% (12,466 MPa vs. 12,615 MPa). For sapwood, relative deviations between batch number 2 and batch number 1 were quite similar for $\overline{E_{m,l,an}}$ (+0.3%, 15,254 MPa vs. 15,209 MPa) and $\overline{E_{min}}$ (14,359 MPa vs. 14,360 MPa). This allowed drawing conclusions on the correct equivalence of the veneer batches in terms of mechanical property. Concerning E_{min}, it was not proven that this indicator allows an accurate prediction of beam bending strength. A simple proportionality trend was shown. The results involving the E_{min} were therefore indicative.

Table 7. Summary results (12% MC).

		Heartwood						Sapwood		
			$E_{m,l,an}$ (MPa)	E_{min} (MPa)				$E_{m,l,an}$ (MPa)	E_{min} (MPa)	
		Amount of Panels × Amount of Beams		Mean Value (COV %)	Mean Value (COV %)	5th percentile (5th p) or Minimum Value (min)	Amount of Panels × Amount of Beams		Mean Value (COV %)	Mean Value (COV %)	5th Percentile (5th p) or Minimum Value (min)
Visual sorting (EN 635-3 standard)	Class II	150 × 5	Random positioning	13,442 (1.3)	12,542 (2.5)	12,008 (5th p)	150 × 5	Random positioning	15,319 (1.4)	14,491 (2.1)	13,962 (5th p)
	Class III-IV	150 × 5	Random positioning	13,541 (1.9)	12,328 (3.5)	11,546 (5th p)	150 × 5	Random positioning	15,122 (2.0)	13,862 (3.4)	12,958 (5th p)
Modulus of elasticity profile sorting	Class A	150 × 5	Random positioning	14,127 (1.4)	13,224 (2.5)	12,615 (5th p)	150 × 5	Random positioning	15,803 (1.2)	14,952 (2.2)	14,360 (5th p)
	Class B	150 × 5	Random positioning	12,889 (1.5)	11,800 (2.9)	11,164 (5th p)	150 × 5	Random positioning	14,599 (1.9)	13,456 (3.0)	12,675 (5th p)
Representative veneer batches	Batch n°1	1 × 5	Favourable optimization	14,472 (0.2)	12,829 (2.4)	12,313 (min)	1 × 5	Favourable optimization	16,205 (0.4)	15,257 (1.8)	14,826 (min)
		150 × 5	Random positioning	13,462 (1.6)	12,615 (2.3)	12,097 (5th p)	150 × 5	Random positioning	15,209 (1.7)	14,360 (2.3)	13,779 (5th p)
	Batch n°2	1 × 4	Favourable optimization	14,413 (0.8)	13,658 (1.3)	13,399 (min)	1 × 5	Unfavourable optimization	14,036 (1.5)	13,091 (1.9)	12,788 (min)
		150 × 5	Random positioning	13,362 (1.8)	12,466 (2.5)	11,908 (5th p)	150 × 5	Random positioning	15,254 (1.9)	14,359 (2.6)	13,712 (5th p)

Table 7 gives information to compare results of favorable and unfavorable optimized beams quantitatively. For sapwood, the favorably optimized beams from batch number 1 had an $\overline{E_{m,l,an}}$ of 16,205 MPa and an $\overline{E_{min}}$ value of 15,257 MPa. It was respectively +15.5% and +16.6% than results of batch number 2 unfavorably optimized beams. The results of batch number 1 optimized beams could also be compared with random beams from the same batch. Thus, relative gaps of +7.6% between optimized beam E_{min} minimum value and random beam E_{min} 5th percentile value (13,779 MPa) were noticed. In terms of modulus of elasticity, the gap between means values was +6.5% (16,205 MPa vs. 15,209 MPa). For heartwood, the favorably optimized beams of batch number 2 had a $\overline{E_{m,l,an}}$ of 14,413 MPa and a $\overline{E_{min}}$ of 13,658 MPa. The batch number 1 heartwood favorably optimized beams had a $\overline{E_{m,l,an}}$ of 14,472 MPa and a $\overline{E_{min}}$ value of 12,829 MPa. This 6.5% difference in $\overline{E_{min}}$ is explained by the fact that, as seen in Figure 6f,g, these beams concentrated excellent local properties at the center of the beams. However, they also had a local minimum in the upper support area that was lower than batch number 2 heartwood favorably optimized beams, mainly due to the no-overlay defect function. For local modulus of elasticity, it is normal to see that this effect had very little impact (-0.4% between 14,413 MPa and 14,472 MPa) compared to the random beams generated from the same veneers.

On the basis of this better homogenization, batch number 2 favorably optimized beams, E_{min} was +12.5% higher than batch number 2 random beams E_{min} 5th percentile value (13,399 MPa vs. 11,908 MPa), while this gap was only +1.8% for heartwood batch number 1 (12,313 MPa vs. 12,097 MPa). For $\overline{E_{m,l,an}}$, theses gaps were respectively +7.9% (14,413 MPa vs. 13,362 MPa) for batch number 2 and +7.5% (14,472 MPa vs. 13,462 MPa) for batch number 1.

As expected, results of the random beams of the four batches of veneer, whether for E_{min} or $E_{m,l,an}$, were always included between Class A and Class B. For example, batch number 1 heartwood random beams $\overline{E_{m,l,an}}$ (13,462 MPa) was between 12,889 MPa of Class B and 14,127 MPa of Class A.

It was also possible to compare the results of optimized beams composed of the veneers representative of the resource with the random beams composed of the strength classes created in [2], i.e., compare the proposed optimized use of the veneer material to the more classical strength. It can be noted that sapwood favorably optimized beams were stiffer in terms of $\overline{E_{m,l,an}}$ than all the random beam classes, and more especially than Class A (+2.5% between 16,205 MPa vs. 15,803 MPa). The same trend was true for E_{min} (a minimum value of 14,826 MPa vs. a 5th percentile value of 14,360 MPa). Similarly, heartwood batch number 2 was superior to all classes ($\overline{E_{m,l,an}}$, 14,413 MPa vs. 14,127 MPa for Class A, E_{min} minimum value of 13,399 MPa vs. E_{min} 5th percentile value of 12,615 MPa for Class A). For heartwood batch number 1, it is true for $\overline{E_{m,l,an}}$ (14,472 MPa vs. 14,127 MPa), but E_{min} minimum value was slightly lower than Class A E_{min} 5th percentile value (12,313 MPa vs. 12,615 MPa). This contributes to show the beneficial effect of avoiding superposition defects, which permitted a better distribution of the defects in the zone between supports and thus improved the E_{min} value in heartwood batch number 2 over heartwood batch number 1. Generally, these results prove that the proposed optimization can allow obtaining better bending stiffness from the full heterogeneous resource of peeled veneers than with a high-quality classical sorting and random LVL composing.

4. Discussion

To the best of authors' knowledge, this study is the first to deal with the optimization of the LVL material by the consideration and exploitation of a highly heterogeneous species, not by the arrangement of the different qualities through the plies, but along the length of the material, according to an edgewise bending testing configuration. This was possible thanks to a new approach based on a numerical model fed with veneer density and local fiber angle in the veneer plan. This last parameter was measured using a new apparatus specifically designed to measure online local veneer fiber orientation.

The effects of the customized were highlighted by favorable or unfavorable optimization and were observed in the analytical and experimental results of the beams of both batches. For tests on the sapwood LVL beams, variations of +11.8% for experimental local modulus of elasticity and +28.1% for bending strength were found on average for favorably optimized batch compared to unfavorable batch.

The results presented in terms of local modulus of elasticity calculation from modelized beams were satisfactory, which means that the prediction of local properties and the analytical calculation are relatively reliable, despite sources of inaccuracies. In terms of pure modeling, the use of veneer global density, neglecting the localized effects of latewood and earlywood, is an important perspective of improvement.

The gain of the favorably optimized beams over random beams was quantifiable thanks to the analytical results, and it was more than 7.5% in terms of stiffness. However, the manufactured beams of this study had a relatively low span. The effectiveness of this method, which consists of coinciding the veneer quality with the curve of the bending moment, remains to be tested in cases of larger beams. Moreover, this method takes advantage of the variability of the veneer resource used by positioning low-quality veneer to low-stressed beam area, and conversely, thus the main advantage is to allow more efficient usage of low-quality resources.

Concerning the use of the nonsuperposition option for heartwood batch number 2, the homogenization effect was not noticeable in experimental results compared to batch number 1. Indeed, in terms of elastic mechanical property, the local modulus averaged the behavior between the loading points of the beam and did not allow for a precise conclusion to be drawn on a phenomenon evolving along this portion. Second, the number of beams was not sufficient to identify trends.

Finally, the modeling of LVL beam strength is a very complex issue. The use of E_{min} as a bending strength indicator did not show very high reliability. It did, however, make it possible to distinguish, on average, the differences in performance between the batches. It was calculated with a very large number of simplifying hypotheses, intrinsically linked to analytical modeling. For example, the conditions and quality of gluing were not taken into account at all. The need to determine an indicator based on other criteria, such as local defect density or average knot spacing, remains.

Author Contributions: Conceptualization, R.D.; methodology, R.D.; software, R.D.; validation, G.P., S.G., L.D.; formal analysis, R.D.; investigation, R.D., G.P., S.G., L.D.; data curation, R.D.; writing—original draft preparation, R.D.; writing—review and editing, R.D., G.P., S.G., L.D.; visualization, R.D.; supervision, G.P., S.G., L.D.; project administration, L.D., G.P.; funding acquisition, G.P. All authors have read and agreed to the published version of the manuscript.

Funding: This study is funded by the region of the Burgundy Franche-Comté, the TreeTrace project subsidized by ANR-17-CE10-0016-03. This study was performed thanks to the partnership built by BOPLI: a shared public-private laboratory built between the Bourgogne Franche-Comté region, LaBoMaP and the Brugère company.

Data Availability Statement: The data presented in this study are available on request from the corresponding author. The data are not publicly available due to industrial application confidentiality.

Acknowledgments: Thanks to Roxane Dodeler for her investment in the experimental task during her Master project period. Thanks to Jean-Claude Butaud for the experimental task preparation and execution.

Conflicts of Interest: The authors declare no conflict of interest. The funders had no role in the design of the study; in the collection, analyses, or interpretation of data; in the writing of the manuscript, or in the decision to publish the results.

References

1. NF EN 635-3. Plywood. Classification by Surface Appearance. Part 3: Softwood. 1 July 1995. Available online: https://sagaweb-afnor-org.rp1.ensam.eu/fr-FR/sw/consultation/notice/1244973?recordfromsearch=True (accessed on 29 October 2020).
2. Duriot, R.; Pot, G.; Girardon, S.; Roux, B.; Marcon, B.; Viguier, J.; Denaud, L. New Perspectives for LVL Manufacturing from Wood of Heterogeneous Quality—Part. 1: Veneer Mechanical Grading Based on Online Local Wood Fiber Orientation Measurement. *Forests* **2021**, *12*, 1264. [CrossRef]
3. Burdurlu, E.; Kilic, M.; İlçe, A.; Uzunkavak, O. The effects of ply organization and loading direction on bending strength and modulus of elasticity in laminated veneer lumber (LVL) obtained from beech (*Fagus orientalis* L.) and lombardy poplar (*Populus nigra* L.). *Constr. Build. Mater.* **2007**, *21*, 1720–1725. [CrossRef]
4. Chen, F.; Deng, J.; Li, X.; Wang, G.; Smith, L.M.; Shi, S.Q. Effect of laminated structure design on the mechanical properties of bamboo-wood hybrid laminated veneer lumber. *Eur. J. Wood Prod.* **2017**, *75*, 439–448. [CrossRef]
5. Chin, K.L.; Tahir, P.M.; H'ng, P.S. Bending Properties of Laminated Veneer Lumber Produced from Keruing (*Dipterocarpus* sp.) Reinforced with Low Density Wood Species. *Asian J. Sci. Res.* **2010**, *3*, 118–125. [CrossRef]
6. NF EN 14374. Structures en bois—LVL (Lamibois)—Exigences. Available online: https://www-batipedia-com.rp1.ensam.eu/pdf/document/QTH.pdf#zoom=100 (accessed on 25 March 2020).
7. NF EN 408+A1. Timber Structures—Structural Timber and Glued Laminated Timber—Determination of Some Physical and Mechanical Properties. 1 September 2012. Available online: https://sagaweb-afnor-org.rp1.ensam.eu/fr-FR/sw/consultation/notice/1401191?recordfromsearch=True (accessed on 29 October 2020).
8. Pollet, C.; Henin, J.-M.; Hébert, J.; Jourez, B. Effect of growth rate on the physical and mechanical properties of Douglas-fir in western Europe. *Can. J. For. Res.* **2017**, *47*, 1056–1065. [CrossRef]
9. Tropix 7. Les principales caractéristiques technologiques de 245 essences forestières tropicales—Douglas. 2012. Available online: https://tropix.cirad.fr/FichiersComplementaires/FR/Temperees/DOUGLAS.pdf (accessed on 29 October 2020).
10. NF-EN-322. Panneaux à base de bois—Determination de l'humidité. 1 June 1993. Available online: https://cobaz-afnor-org.rp1.ensam.eu/notice/norme/nf-en-322/FA020887?rechercheID=1757547&searchIndex=1&activeTab=all#id_lang_1_descripteur (accessed on 8 June 2021).
11. NF EN 384+A1. Structural Timber—Determination of Characteristic Values of Mechanical Properties and Density. 21 November 2018. Available online: https://sagaweb-afnor-org.rp1.ensam.eu/fr-FR/sw/consultation/notice/1539784?recordfromsearch=True (accessed on 29 October 2020).
12. Woodtec Fankhauser GmbH. Vacuum Press. Available online: https://www.woodtec.ch/wp-content/uploads/2020/11/Documentation-Vacuum-Press.pdf (accessed on 31 March 2021).
13. Viguier, J.; Bourgeay, C.; Rohumaa, A.; Pot, G.; Denaud, L. An innovative method based on grain angle measurement to sort veneer and predict mechanical properties of beech laminated veneer lumber. *Constr. Build. Mater.* **2018**, *181*, 146–155. [CrossRef]
14. Olsson, A.; Oscarsson, J.; Serrano, E.; Källsner, B.; Johansson, M.; Enquist, B. Prediction of timber bending strength and in-member cross-sectional stiffness variation on the basis of local wood fibre orientation. *Eur. J. Wood Wood Prod.* **2013**, *71*, 319–333. [CrossRef]
15. Viguier, J.; Bourreau, D.; Bocquet, J.-F.; Pot, G.; Bléron, L.; Lanvin, J.-D. Modelling mechanical properties of spruce and Douglas fir timber by means of X-ray and grain angle measurements for strength grading purpose. *Eur. J. Wood Wood Prod.* **2017**, *75*, 527–541. [CrossRef]

Article

Penetration of Different Liquids in Wood-Based Composites: The Effect of Adsorption Energy

Hamid R. Taghiyari [1], Roya Majidi [2], Mahnaz Ghezel Arsalan [1], Asaad Moradiyan [1], Holger Militz [3], George Ntalos [4] and Antonios N. Papadopoulos [5,*]

[1] Wood Science and Technology Department, Faculty of Materials Engineering & New Technologies, Shahid Rajaee Teacher Training University, Tehran 16788-15811, Iran; htaghiyari@sru.ac.ir (H.R.T.); mahnaz_arsalan2@yahoo.com (M.G.A.); pranawood66@gmail.com (A.M.)

[2] Department of Physics, Faculty of Sciences, ShahidRajaee Teacher Training University, Tehran 16788-15811, Iran; r.majidi@sru.ac.ir

[3] Wood Biology and Wood Products, Georg-August-University Göttingen, 37077 Göttingen, Germany; hmlitz@gwdg.de

[4] Department of Forestry, Wood Science and Design, University of Thessaly, GR-431 00 Karditsa, Greece; gnatalos@uth.gr

[5] Laboratory of Wood Chemistry and Technology, Department of Forestry and Natural Environment, International Hellenic University, GR-661 00 Drama, Greece

* Correspondence: antpap@for.ihu.gr; Tel.: +30-252-106-0445

Citation: Taghiyari, H.R.; Majidi, R.; Arsalan, M.G.; Moradiyan, A.; Militz, H.; Ntalos, G.; Papadopoulos, A.N. Penetration of Different Liquids in Wood-Based Composites: The Effect of Adsorption Energy. *Forests* **2021**, *12*, 63. https://doi.org/10.3390/f12010063

Received: 14 December 2020

Accepted: 4 January 2021

Published: 7 January 2021

Abstract: The penetration properties of three different liquids on the surface of medium-density fiberboard (MDF) and particleboard panels were studied. Water, as a polar liquid, was compared to two other less polar liquids (namely, ethanol and kerosene) with significantly larger molecules. Measurement of penetration time and wetted area demonstrated significantly higher values for water in comparison with the other two liquids, in both composite types. Calculation of adsorption energies, as well as adsorption distances, of the three liquid molecules on hemicellulose showed higher potentiality of water molecules in forming bonds on hemicellulose. However, comparison of the adsorption energies of cellulose with hemicellulose indicated a higher impact of the formation of bonds between hydroxyl groups in water and cellulose in hindering the penetration of water molecules into the composite textures. It was concluded that the formation of strong and stable bonds between the hydroxyl groups in water and cellulose resulted in a significant increase in penetration time and wetted area.

Keywords: adsorption properties; hydroxyl groups; medium-density fiberboard (MDF); particleboard panels; polar molecules

1. Introduction

Wood-based composite panels with desirable length and width can satisfy the growing needs in the home and office furniture market; therefore, these panels have become an essential part of household and office lives in many countries [1–9]. However, usually, the wood-based composite panels are susceptible to water and moist conditions, particularly those panels produced with urea-formaldehyde (UF) resin [10–12].

Concerning the description of wetting processes, and the prediction of the wettability of a solid by different liquids, the surface tensions of both the liquid and solid are of importance. All thermodynamic theories concerning the wetting behavior are based on a flat and isotropic (ideal) surface of the solid, which, however, does not exist in practice. For measuring contact angles on real surfaces, it is important to know the various parameters influencing the contact angles of liquid droplets applied onto the surfaces. These parameters depend on the nature and properties of the wetting liquid as well as of the wetted solid surface. This especially is an unalterable presupposition for measuring contact angles on wood, which differs from many other materials in terms of its cellular

91

structure. One point to consider is the effect of the roughness of the wood surface, including the influence of the wood moisture content of the samples on the dynamic contact angles and on the surface tensions of solid wood, calculated from these contact angles. Zorll [13] infers that liquids still are able to wet such a rough surface, whereas on an ideal flat surface with no roughness, no proper wetting occurs. Wettability measurements performed by Hse [14] using phenol-formaldehyde resins and pine veneers showed lower static contact angles on the rougher early-wood surfaces than on the smoother latewood surfaces.

As wood is a hygroscopic material, it must be assumed that the moisture content of wood influences the wettability of wood surfaces by liquids. When a droplet of water comes into contact with dry wood, swelling processes of wood take place underneath the droplet and the wood surface structure might change. A low wood moisture content impedes the wettability. A possible reason for this can be the lower number of hydrophilic hydroxy groups of the wood, which act as a sorption place for water molecules [15]. This consideration, however, is only valid for wetting wood with aqueous liquids or alcohols. Severe conditions, e.g., high temperatures during hot pressing in the production of wood-based panels, might cause a decrease of the number of cellulose-based OH-groups with a lower wood moisture content, which causes a thermic deactivation of the wood surface. Upon applying heat energy to the surface, an activation of wood components might occur, causing cleavage of chemical bonds [16].

In this connection, having a better outlook on the penetration patterns of different liquids that may either affect wood or the resin used in the composite would provide both researchers and the industry sector with an in-depth understanding of how the panels may be affected by different liquids. Therefore, the present study was designed to find out how water, as a polar liquid, and two other less polar liquids (namely, ethanol and kerosene) penetrate into and expand throughout the surface of composite panels. Two popular wood-based composites were chosen for this purpose: medium-density fiberboard (MDF) and particleboard panels. Urea-formaldehyde resin was used as a binder in both panels, however, further studies on panels with other binders (like phenol-formaldehyde and melamine-formaldehyde resins) would provide additional useful information for comparison purposes. Moreover, the adsorption energy and adsorption distance of the above mentioned liquids on hemicellulose were also calculated by density functional theory (DFT) to further investigate to what extent theoretical studies comply with experimental tests.

2. Materials and Methods

2.1. Specimen Preparation

Three MDF panels and three particleboard panels from industrial production were selected. MDF panels were produced by Khazar Company, Amol, Iran, with a density of 750 kg.m^{-3}. Urea-formaldehyde resin content was 12% based on the dry weight of wood fibers, with an addition of 0.5% liquid paraffin. Fibers were comprised of poplar (*Populus nigra*) (50–60%), a mixture of industrial wood species (30–40%, including beech *Fagus orientalis*, alder *Alnus glutinosa*, and hornbeam *Carpinus betulus*, with rather similar proportions), and less than 5% pruned branches from fruit trees. All trees were cut from the neighboring forests located in Amol (Iran). These species were chosen by Khazar Co. to be used in their production program based on the availability in the region. The pH of the resin was 7. Continuous hot-press was used with a temperature range of 150–260 °C at different sections of the press. The duration of hot-press was 2 min. Sanding was carried out on both sides of all panels, using sanding papers with 80, 100, and 150 grades. About 0.35–0.40 mm was sanded from each side of the panels. Particleboard panels were produced by Iran-Choob Company located in Ghazvin Industrial Complex, Ghazvin, Iran, the density of which was 730 kg.m^{-3}. Wood chips consisted poplar (*Populus nigra*) (80–85%) grown in the region and pruned branches from the neighboring fruit gardens (10–15%). These species and raw materials were chosen by the producing factory based on the fact that poplar is considered a low density species, which makes it favorable for particleboard production. Moreover, the availability of poplar and pruned branches was of importance.

Urea-formaldehyde resin contents were 17.5% and 10% based on the dry weight of wood chips for the surface and core layers, respectively. An addition of 1% ammonium chloride was used as a hardener. The pH of the resin was 7–8. A fixed hot-press was used with a temperature of 200 °C. The specific pressure of the plates was 24 kg/cm^2 (with 200 kgf as the total nominal pressure). The duration of the hot-press was 3 min. Sanding was carried out on both sides of all panels, using sanding papers with 50, 100, and 120 grades. About 0.40–0.50 mm was sanded from each side of the panels. Based on the information provided by the seller, both composite panel types used urea-formaldehyde resin as a binder. The thickness of all panels was 16 mm. From the center and two corners of each panel, two boards were cut with dimensions of 30 cm × 30 cm. From each board, three specimens were cut, the dimensions of which were 5 cm × 5 cm. In total, 27 specimens were prepared from either of the composite panels to be divided by three for each of the liquids (water, ethanol, and kerosene).

2.2. Measurement of Penetration Time

A drop of liquid was put on the surface center of each specimen (MDF or particleboard) by a 10 mL lab glass division dropper pipette. The time from the moment the drop first touched the specimen surface until it fully penetrated and disappeared was measured as the penetration time. The whole process was monitored by a 10× light magnifier.

2.3. Measurement of Wetted Area

Once the drop of liquid (water, ethanol, or kerosene) penetrated into the surface of the specimen and disappeared fully, the specimen was put on a stand to take a photo from the wetted spot at a constant distance between the surface and lens for all specimens. This confirmed that the magnification for all specimens was the same. Each and every photo was first converted to its negative form with black-and-white colour (Figure 1), and then analyzed by ImageJ software (version 1.53a, Wisconsin, WI, USA), to measure the wetted area. The area was then converted to square millimeter for comparison and reporting purposes.

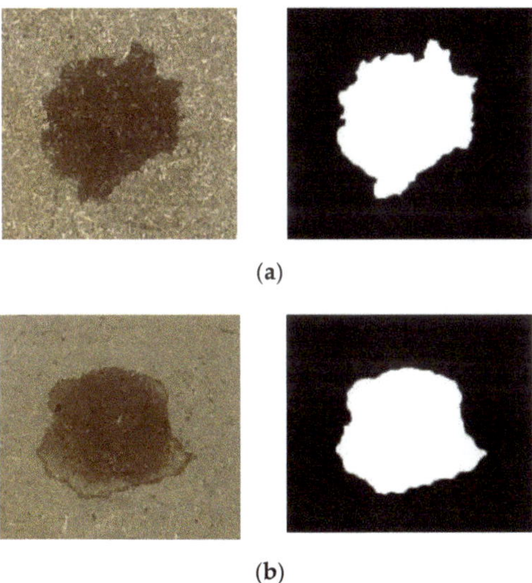

(a)

(b)

Figure 1. Photo at a constant distance from the wetted spot and the counterpart negative image of the same specimen in black-and-white in particleboard (**a**) and medium-density fiberboard (**b**) specimens.

2.4. Density Functional Theory (DFT)

DFT simulations were carried out as instructed in the Open MX3.8 package [17]. The generalized gradient approximation, done with the aid of Perdew–Burke–Ernzerhof correction, was employed in order to get to the exchange-correlation function [18]. The DFT-D2 approach was used to include the long-range van der Waals interactions [19]. The plane wave cut-off energy was fixed at 60 Ry. The model of hemi-cellulose introduced by Kaith et al. [20] was considered in this study (Figure 2). All geometries were optimized until the maximum force on each atom was smaller than 0.001 eV/Å.

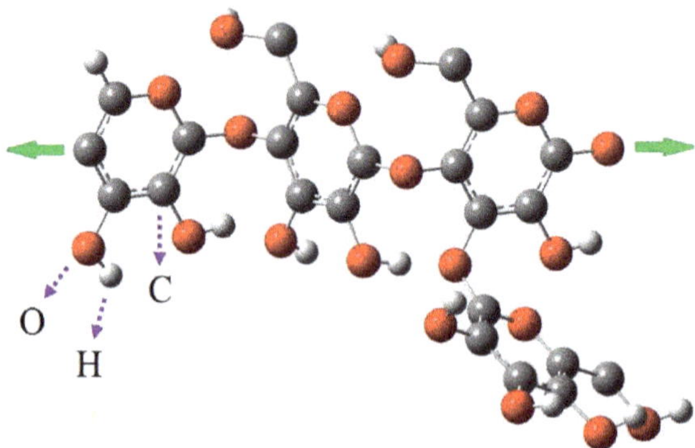

Figure 2. Atomic structure of hemi-cellulose.

In order to better interpret the potential interactions between each of the three liquids (water, ethanol, and kerosene), adsorption distances and adsorption energies of these three molecules on the hemicellulose surface were calculated by DFT. Each of the three molecules of water (H_2O), ethanol (C_2H_5OH), or kerosene ($C_{12}H_{26}$) were rotated on the hemicellulose surface to achieve the best configuration. Each process was initiated with the farthest distance, and then the distance became shortest until the optimal adsorption distance was achieved. Adsorption energy was calculated by Equation (1).

$$E_{ads} = E_{hemicellulose+molecule} - (E_{hemicellulose} + E_{molecule}) \tag{1}$$

where $E_{hemicellulose+molecule}$ represents the total energy of hemicellulose including energy of the adsorbed molecule, and $E_{hemicellulose}$ and $E_{molecule}$ are the total energies of the hemicellulose surface and each of the adsorbed molecules (water, ethanol, and kerosene), respectively. The negative adsorption energy corresponded to the structure with the highest stability. Therefore, the highest stable construction was chosen as the ultimate configuration.

2.5. Statistical Analysis

The two composite panel types (MDF and particleboard) were produced and processed under different conditions, including the size of the materials (fibers and chips), resin content, surface sanding, tree species, and hot-press temperature. Therefore, a comparison between the two types of panels from a statistical point of view was not logical, because of the multiple sources of variance and factors involved. Therefore, analysis of variance (ANOVA) was separately carried out for the data and mean values of each composite type. However, both panel types are presented in joint graphs so that the readers would have a better outlook on the overall trends between the properties measured on the two types of composite panels. The calculations were carried out using SPSS/18 (2010).

3. Results

3.1. Penetration Time

The results of the penetration of one drop of each of the three kinds of liquids demonstrated that the highest and lowest penetration times were found in MDF-water (2918.7 s ± 507) and MDF-alcohol (10.2 s ± 3), respectively. Penetration of a water drop took significantly longer with water in comparison with the other two less polar liquids (alcohol and kerosene),in both MDF and particleboard specimens (Figure 3). No statistically significant difference was observed between the penetration of alcohol and kerosene, though fluctuations were observed between these liquids in the two composite panels.

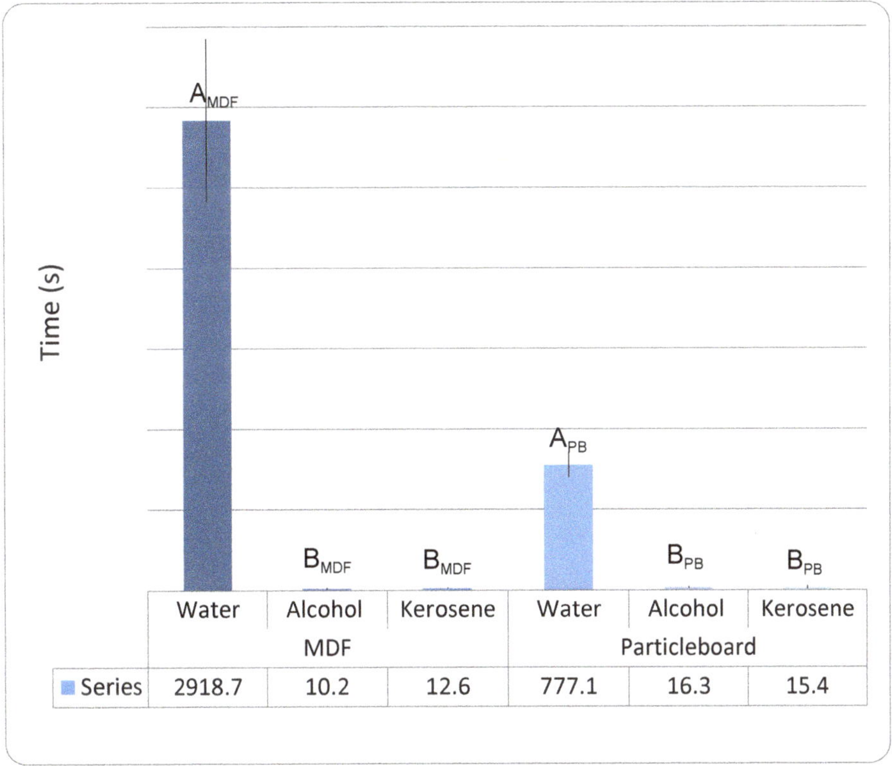

	Water	Alcohol	Kerosene	Water	Alcohol	Kerosene
		MDF			Particleboard	
■ Series	2918.7	10.2	12.6	777.1	16.3	15.4

Figure 3. Penetration time for one drop of each of the three liquids (water, ethanol, and kerosene) to fully disappear from the surface of wood-based composite specimens (letters on each column represent Duncan's multiple range test, $\alpha = 0.05$, separately carried out for either MDF or particleboard). MDF = medium-density fiberboard; PB = particleboard.

3.2. Wetted Area

The results of the measurement of the wetted area illustrated that water made a wider wetted area in comparison with alcohol and kerosene, in both MDF and particleboard composite panels. Unlike penetration time, a significant difference was observed between alcohol and kerosene in wetting the surface areas of both composite types (Figure 4). Wetted areas in particleboard were larger in comparison with their counterparts in MDF specimens for all three liquids.

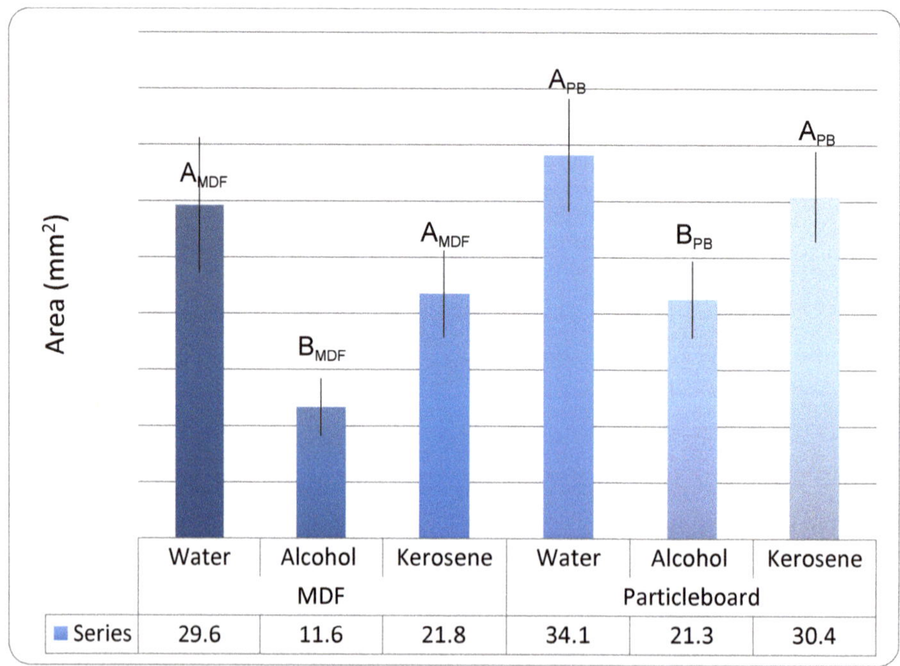

Figure 4. The wetted area in the wood-based composite specimens measured once the drop of liquid penetrated into the surface of composite specimen and fully disappeared (letters on each column represent Duncan's multiple range test, $\alpha = 0.05$, separately carried out for either MDF or particleboard). MDF = medium-density fiberboard; PB = particleboard.

3.3. Density Functional Theory (DFT)

The adsorption energies between hemicellulose and the three liquids (water, ethanol, and kerosene), as well as the nearest adsorption distances, are listed in Table 1. The calculation of adsorption energy of the liquids on hemicellulose demonstrated that all three liquids were adsorbed on the hemicellulose surface. The nearest adsorption distance (1.39 Å) and the strongest adsorption energy (-1.16 eV) were found in water molecules. As the smallest molecule between the three liquids, the model illustrated that water had the potential of penetrating into the hemicellulose complex and branched structure (Figure 5).

Table 1. Adsorption distance and adsorption energy for water, ethanol, and kerosene on hemicellulose.

Adsorption Distance and Energy	Water $OH_w \dots O$	Ethyl Alcohol $OH_e \dots O$	Kerosene $CH \dots C_k$
d (Å)	1.39	1.40	1.67
E_{ads} (eV)	-1.16	-0.94	-1.05

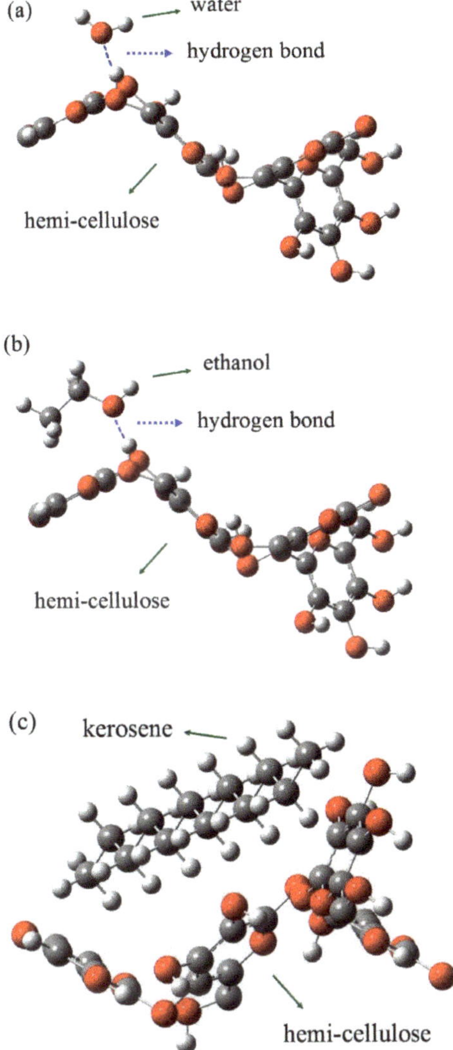

Figure 5. Schematic view of water (H_2O) (**a**), ethanol (C_2H_5OH) (**b**), and kerosene ($C_{12}H_{26}$) (**c**) molecules adsorbed on hemicellulose (the hydrogen bonds are shown with a dashed line).

4. Discussion

The highest adsorption energy, as well as the shortest adsorption distance, were both found between water molecules and the hemicellulose surface. The high adsorption energy translated to the formation of new bonds between water molecules and hemicellulose, eventually leading to a more stable bond between water molecules and hemicellulose. This partially accounted for slower movement and transfer of water molecules through the voids and spaces in both composite panels, resulting in a significant increase in penetration time for water in comparison with ethanol and kerosene (Figure 3). Though not statistically significant, Pearson correlations between the penetration time and wetted area calculated for the three liquids illustrated that correlation values for water were negative (in both MDF and particleboard specimens), while correlation values for ethanol and kerosene were positive. In this connection, a previous study on the liquid permeability of different medium-density fiberboard panels similarly reported that all MDF treatments indicated significantly lower water permeability in comparison with alcohol and kerosene permeability values [21]. The cited authors explained that water molecules primarily bonded with the hydroxyl groups of cellulose and hemicellulose in the cell wall [21–24]. These bonds delayed the movement and transfer of water molecules through the continuous voids and spaces in the composite specimens. However, the difference between the adsorption energies and distances of the three liquids on hemicellulose were not as great as the difference of the three liquids on cellulose, nor were they compatible with the difference between the penetration time or liquid permeability of the three liquids.

Further comparison between the adsorption energies of the three liquids (water, ethanol, and kerosene) on cellulose and hemicellulose revealed a significant difference between the energies of cellulose and hemicellulose, though the results of both the present and previous studies [21] clearly confirmed that water molecules have a higher degree of interaction with the hydroxyl groups of cell wall polymers (cellulose and hemicellulose) in comparison with ethanol and kerosene. That is, the adsorption energy of water molecules on cellulose was 55% and 44% higher in comparison with ethanol and kerosene, respectively [21]. However, the adsorption energy of water molecules on hemicellulose was only 19% and 9% higher in comparison with the counterpart energies of ethanol and kerosene on hemicellulose, respectively. This indicated that the significant differences between the penetration time and wetted area, as well as liquid permeability as studied previously [21], were mostly attributed to the difference of adsorption energy of the three liquids on cellulose rather than the low difference observed on hemicellulose. In this connection, further studies may evaluate the penetration time and wetted area in wood-based composite panels made with other types of resins than urea-formaldehyde resin to find out how different resins may affect the liquid penetration properties.

5. Conclusions

Penetration of water, a polar liquid with rather small molecules, into two wood-based composites was compared with two other less polar liquids, namely ethanol and kerosene. Full penetration of one drop of each of the three liquids into the composite texture was measured as the penetration time, and the area into which the drop of liquid expanded on the surface of the specimens was also measured and reported as the wetted area. Water demonstrated a significantly higher penetration time and wetted area. Density functional theory indicated a higher tendency of water molecules in forming more stable bonds on hemicellulose. Comparison of the three liquids on cellulose and hemicellulose demonstrated a higher impact of bonds between hydroxyl groups in water and cellulose in comparison with the bonds formed between water and hemicellulose. It was concluded that the higher penetration time and wetted area in water, in comparison with ethanol and kerosene, were mainly attributed to the formation of stronger and more stable bonds between the hydroxyl groups in water with cellulose.

Author Contributions: Methodology, H.R.T.; Validation, H.R.T., R.M., H.M., G.N., and A.N.P.; Investigation, H.R.T., M.G.A., and A.M.; Writing—Original Draft Preparation, H.R.T., R.M., and A.N.P.; Writing—Review and Editing, H.R.T., H.M., and A.N.P.; Visualization, H.R.T. and R.M.; Supervision, H.R.T. and A.N.P. All authors have read and agreed to the published version of the manuscript.

Funding: This research received no external funding.

Institutional Review Board Statement: Not applicable.

Informed Consent Statement: Not applicable.

Data Availability Statement: The data presented in this study are available on request from the corresponding author.

Acknowledgments: The first author appreciates constant scientific support of Jack Norton (Retired, Horticulture & Forestry Science, Queensland Department of Agriculture and Fisheries, Australia), as well as Alexander von Humboldt Stiftung, Germany.

Conflicts of Interest: The authors declare no conflict of interest.

References

1. Taghiyari, H.R.; Majidi, R.; Esmailpour, A.; Sarvari Samadi, Y.; Jahangiri, A.; Papadopoulos, A.N. Engineering composites made from wood and chicken feather bonded with UF resin fortified with wollastonite: A novel approach. *Polymers* **2020**, *12*, 857. [CrossRef] [PubMed]
2. Papadopoulos, A.N.; Gkaraveli, A. Dimensional stabilization and strength of particleboard by chemical modification with propionic anhydride. *HolzalsRoh- Werkst.* **2003**, *61*, 142–144. [CrossRef]
3. Goodge, K.; Frey, M. Biotin-conjugated cellulose nanofibers prepared via copper-catalyzed alkyne-azide cycloaddition (CuAAC) "click" chemistry. *Nanomaterials* **2020**, *10*, 1172. [CrossRef] [PubMed]
4. Taghiyari, H.R.; Esmailpour, A.; Majidi, R.; Hassani, V.; Abdolah Mirzaei, R.; Farajpour Bibalan, O.; Papadopoulos, A.N. The effect of silver and copper nanoparticles as resin fillers on less-studies properties of UF-based particleboards. *Wood Mater. Sci. Eng.* **2020**. [CrossRef]
5. Platnieks, O.; Gaidukovs, S.; Barkane, A.; Sereda, A.; Gaidukova, G.; Grase, L.; Thakur, V.K.; Filipova, I.; Fridrihsone, V.; Skute, M.; et al. Bio-Based Poly(butylene succinate)/Microcrystalline Cellulose/Nanofibrillated Cellulose-Based Sustainable Polymer Composites: Thermo-Mechanical and Biodegradation Studies. *Polymers* **2020**, *12*, 1472. [CrossRef] [PubMed]
6. Shaoqiu, K.; Wang, Z.; Zhang, K.; Cheng, F.; Sun, J.; Wang, N.; Zhu, Y. Flexible conductive cellulose network-based composite hydrogel for multifunctional supercapacitors. *Polymers* **2020**, *12*, 1369.
7. Hassani, V.; Papadopoulos, A.N.; Schmidt, O.; Maleki, S.; Papadopoulos, A.N. Mechanical and Physical Properties of Oriented Strand Lumber (OSL): The Effect of Fortification Level of Nanowollastonite on UF Resin. *Polymers* **2019**, *11*, 1884. [CrossRef]
8. Taghiyari, H.R.; Esmailpour, A.; Majidi, R.; Morrell, J.J.; Mallaki, M.; Militz, H.; Papadopoulos, A.N. Potential Use of Wollastonite as a Filler in UF Resin Based Medium-Density Fiberboard (MDF). *Polymers* **2020**, *12*, 1435. [CrossRef]
9. Papadopoulos, A.N.; Taghiyari, H.R. Innovative wood surface treatments based on nanotechnology. *Coatings* **2019**, *9*, 866. [CrossRef]
10. Pizzi, A.; Papadopoulos, A.N.; Policardi, F. Wood composites and their polymer binders. *Polymers* **2020**, *12*, 1115. [CrossRef] [PubMed]
11. Wang, X.; Chen, X.; Xie, X.; Yuan, Z.; Cai, S.; Li, Y. Effect of phenol formaldehyde resin penetration on the quasi-static and dynamic mechanics of wood cell walls using nanoindentation. *Nanomaterials* **2019**, *9*, 1409. [CrossRef] [PubMed]
12. Antov, P.; Savov, V.; Mantanis, G.; Neykov, N. Medium-density fibreboards bonded with phenol-formaldehyde resin and calcium lignosulfonate as an eco-friendly additive. *Wood Mat. Sci. Eng.* **2020**. [CrossRef]
13. Zorll, U. Neue Erkenntnisseüber die Bedeutung der Benetzungfür die AdhäsionbeiBeschichtungs- und Klebstoffen. *Adhäsion* **1978**, *22*, 320–325.
14. Hse, C.Y. Wettability of southern pine veneer by phenol formaldehyde wood adhesives. *For. Prod. J.* **1972**, *22*, 51–56.
15. Scheikl, M.; Dunky, M. Measurement of dynamic and static contact angles on wood for the determination of its surface tension and the penetration of liquids into the wood surface. *Holzforschung* **1998**, *52*, 89–94. [CrossRef]
16. Scheikl, M.; Dunky, M. Softwareunterstütztestatische und dynamischeKontaktwinkelmeßmethodenbei der Benetzung von Holz. *HolzalsRoh- Werkst.* **1996**, *54*, 113–117. [CrossRef]
17. Ozaki, T.; Kino, H.; Yu, J.; Han, M.J. User's Manual of OpenMX Version 3.8. 2018. Available online: http://www.openmx-square.org/openmx_man3.8/openmx.html (accessed on 20 August 2018).
18. Perdew, J.P.; Burke, K.; Ernzerhof, M. Generalized Gradient Approximation Made Simple. *Phys. Rev. Lett.* **1996**, *77*, 3865. [CrossRef] [PubMed]
19. Grimme, S. Semiempirical GGA-type density functional constructed with a long-range dispersion correction. *J. Comput. Chem.* **2006**, *27*, 1787–1799. [CrossRef] [PubMed]
20. Kaith, B.S.; Mittal, H.; Jindal, R.; Maiti, M.; Kalia, S. Environment Benevolent Biodegradable Polymers: Synthesis, Biodegradability, and Applications. In *Cellulose Fibers: Bio- and Nano-Polymer Composites*; Kalia, S., Kaith, B., Kaur, I., Eds.; Springer: Berlin/Heidelberg, Germany, 2011.
21. Esmailpour, A.; Taghiyari, H.R.; Majidi, R.; Babaali, S.; Morrell, J.J.; Mohammadpanah, B. Effects of adsorption energy on air and liquid permeability of nanowollastonite-treated medium-density fiberboard. *IEEE Trans. Instrum. Meas.* **2020**. [CrossRef]
22. Siau, J.F. *Wood: Influence of Moisture on Physical Properties*; Department of Wood Science and Forest Products Virginian Polytechnic Institute and State University: Blacksburg, VA, USA, 1995.
23. Siau, J.F. *Transport Processes in Wood*; Springer-Verlag: Berlin/Heidelberg, Germany; GmbH & Co. KG: Berlin, Germany, 2011.
24. Skaar, C. *Wood-Water Relations*; Springer: Berlin/Heidelberg, Germany, 1988.

 *forests*

Article

The Effect of Veneer Densification Temperature and Wood Species on the Plywood Properties Made from Alternate Layers of Densified and Non-Densified Veneers

Emilia-Adela Salca, Pavlo Bekhta and Yaroslav Seblii

1   Faculty of Wood Engineering, Transilvania University of Brasov, Eroilor 29, 500036 Brasov, Romania
2   Department of Wood-Based Composites, Cellulose and Paper, Ukrainian National Forestry University, Gen. Chuprynky 103, 79057 Lviv, Ukraine; yruk1996@gmail.com
*   Correspondence: emilia.salca@unitbv.ro (E.-A.S.); bekhta@nltu.edu.ua (P.B.)

Received: 25 May 2020; Accepted: 23 June 2020; Published: 24 June 2020

Abstract: In this study the properties of plywood manufactured from densified and non-densified veneer sheets and alternate layers of such veneers with and without densification using low amount of adhesive as a function of densification temperature and wood species were investigated. The plywood panels were made from rotary-cut birch and black alder veneers using urea-formaldehyde (UF) adhesive. Veneer sheets with thickness of 1.5 mm were subjected to the thermal-compression at three different temperatures while keeping constant the pressure during a same time span. Five-layers plywood panels were produced using a constant hot-pressing schedule using different amounts of glue spread as a function of the plywood type; such as plywood made from non-densified (80 g/m^2) and densified (60 g/m^2) veneers only; and combination of them (70 g/m^2). The bending strength (MOR) and the modulus of elasticity (MOE) along with the shear strength of the plywood samples for bonding class 1 (dry conditions) have been determined. As expected bending strength of the plywood samples increased with the increasing in density. The increase of veneer densification temperature resulted in a gradually decrease of MOR; MOE and shear strength values for the plywood panels made of densified veneers and mixed panels of both species. The temperature of 150 °C for veneer densification seemed to be enough to achieve enhanced bending and bonding properties. All plywood panels in this study were manufactured using reduced glue consumption and they presented satisfactory properties performance for indoor applications.

Keywords: birch; black alder; densification; plywood properties

1. Introduction

Plywood has an attractive appearance of natural wood as compared to that of other wood-based composite panels. It is mostly used for furniture manufacturing and indoor design but also it can be used for interior and exterior building production [1,2]. The strength properties of the product along with its dimensional stability are required in such sectors. Several factors have major influence on the mechanical properties of plywood, especially on the bending strength (MOR) and modulus of elasticity (MOE), namely wood species, density, structure, moisture content, number of veneer layers, and type of adhesive and its spreading [2–7]. Various modification methods can be used to improve the properties of such products, the thermo-mechanical densification at elevated temperatures of solid wood is one of these methods [8–10].

It is also fact that plywood produced from compressed veneers sheets has enhanced properties [11,12]. The combination of heat and compression is an environmentally friendly method and influences overall the glue line strength of the final product [13]. But a special attention should be granted to the densification because an excessive process may cause negative effects, such as heavier plywood and higher dimensional changes in certain conditions and lower glue bond performance [1,7]. There is a high interest to manufacture environmentally friendly engineering products by using less adhesive spread rate [1]. Studies in this field showed that the veneer densification applied prior to plywood manufacturing reduced (by 40%) the glue spread and pressure while the shear strength of the finished product was improved [14–16].

In previous studies, the veneers of one [14,16–19] or two thicknesses [1] were used in thermo-mechanical densification process to produce plywood mainly with non-densified or densified veneers, or with densified outer layers and non-densified core.

There is little or no information on the properties of plywood when manufactured from combinations of veneers with and without densification in one plywood structure. Therefore the aim of this work was to evaluate the effect of veneer densification temperature and wood species on the plywood properties made from alternate layers of densified and non-densified veneers. The combinations of densified and non-densified veneers within the same plywood structure are expected to generate value-added wood-based products.

2. Materials and Methods

Commercially manufactured rotary-cut veneers of birch (*Betula verrucosa* Ehrh.) and black alder (*Alnus glutinosa* L.) native in Ukraine were used for the experiments. The veneer sheets had a thickness of 1.5 mm and a moisture content of 7.4%. A total of 140 veneer samples with dimensions of 300×300 mm^2 were prepared for the experiments.

2.1. Veneer Densification

The densification process of the veneers was performed under laboratory conditions. A constant pressure of 2 MPa for 3 min time span was applied at three different temperature levels of 150 °C, 180 °C and 210 °C, respectively. The moisture content of densified veneer was of about 2.2%.

2.2. Plywood Preparation

Five-layer plywood samples were manufactured from the two species under laboratory conditions. Three types of plywood from each species were produced by employing same press parameters including pressure, temperature and time. The glue spread varied as function of the plywood type made of non-densified veneers (N), densified veneers (D) and mixed (M) as a combination of densified (d) and non-densified veneers (n) within the same structure (d-n-d-n-d) as it is displayed in Table 1.

Table 1. The pressing parameters for plywood manufacturing and plywood structures.

Wood Species	Type of Plywood per Species	Codification: Densification Temperature and Plywood Symbol		Glue Spread, g/m^2	Pressure, MPa	Temperature, °C	Pressing Time, s
				Variable		Constant Schedule	
					1.8	130	270 + 60
Black alder/ Birch	Plywood made of non-densified veneers	0	N	80			
	Plywood made of densified veneers and densification temperature	150 180 210	D	60			
	Plywood made of densified and non-densified veneers and densification temperature	150 180 210	M	70			

Usually in practice, under industrial conditions, the glue spread ranges from 110 to 150 g/m^2, as it is recommended by the adhesive producer. But in this study the adhesive ratios were reduced almost 50% of the recommended value. Two plywood panels were manufactured for each one of the plywood structure type.

The pressure was continuously reduced to 0 MPa for the last 60 s of the pressing cycle. To manufacture the plywood a commercial urea-formaldehyde (UF) resin was used (specific gravity = 1.28 g/m^3, solid content = 65 ± 2%, pH = 7.5 ± 0.5, viscosity 1000–2500 mPa s at 20 °C). The adhesive was prepared using UF resin, caolin (as a filler), 20% solution of NH$_4$Cl (as a hardener) and distilled water. The adhesive was manually applied onto one side of each uneven veneer sheet by using a roller. The veneers were perpendicularly oriented, layer by layer, in each plywood structure. Prior to any process step and test, both veneers with and without pre-treatment and the produced plywood structures were kept under constant room conditions at 20 ± 2 °C and 65 ± 5% relative humidity for a week.

2.3. Density and Compression Ratio of the Plywood Samples

The density of all types of plywood was determined according to EN 323 standard [20]. The compression ratio CR$_p$ of the plywood panels was calculated as shown below:

$$CR_p = (T_v - T_p)/T_v \times 100 \ (\%) \tag{1}$$

in which T$_v$ is the total thickness of all veneers (mm) and T$_p$ is the thickness of the panel.

2.4. Bending Strength and Modulus of Elasticity of the Plywood Samples

The bending strength (MOR) and the modulus of elasticity (MOE) have been determined according to EN 310 standard for each plywood type [21]. To test the mechanical properties the universal testing machine of SEIDNER type (Form + Test SEIDNER D-7940 Riedlingen West Germany) was used. The reported results of bending properties represent the average of ten samples for each plywood type.

2.5. Shear Strength of the Plywood Samples

The shear strength of the plywood samples was measured according to EN 314-1 [22] and EN 314-2 [23] standards for the plywood bonding class 1 (dry conditions). The samples were immersed in water at 20 ± 3 °C for 24 h. The reported shear strength data represent the average of 20 samples for each plywood batch type.

2.6. Processing of Data

All the data obtained in this study were statistically processed by employing the Minitab 17.3.1 software. Multifactorial ANOVA analysis of variance (95% CI) for the mechanical properties and the response optimization between species for the best strength values of plywood samples were applied.

3. Results and Discussions

3.1. Thickness, Compression Ratio, and Density of the Plywood Samples

The density values along with the thickness and compression ratio of the plywood panels produced in this study are presented in Table 2. Based on the density differences between the two wood species, the density values for all the plywood categories made of birch veneers are higher than the values for black alder plywood panels. The values of plywood thickness for all types of panels were lower than the permissible values stipulated by EN 315 for thin veneers, in the range of 5.66–7.24 mm [24].

Table 2. Thickness, compression ratio, and density of the plywood samples.

Densification Temperature and Type of Plywood		0 N	150 D	180 D	210 D	150 M	180 M	210 M
Black alder	Thickness, mm	6.98 (0.18) *	6.84 (0.11)	6.74 (0.05)	6.48 (0.20)	6.80 (0.06)	6.79 (0.02)	6.50 (0.14)
	Compression ratio, %	11.1	5.0	3.7	0.3	8.8	7.5	7.7
	Density, kg/m^3	607.4 (23.2)	608.9 (28.5)	619.9 (2.1)	637.3 (19.1)	607.1 (23.8)	615.0 (21.1)	623.8 (14.5)
Birch	Thickness, mm	7.03 (0.05)	6.96 (0.04)	6.76 (0.13)	6.72 (0.13)	7.06 (0.15)	6.86 (0.12)	6.83 (0.07)
	Compression ratio, %	10.4	3.3	3.4	1.2	5.4	6.5	5.4
	Density, kg/m^3	747.0 (20.5)	807.9 (6.4)	809.3 (13.8)	828.2 (7.1)	766.8 (4.7)	769.8 (31.1)	805.3 (27.4)

* numbers in parenthesis are standard deviations values.

3.2. Bending Properties of the Plywod Samples

The bending strength and the modulus of elasticity presented higher average values for birch plywood compared to black alder plywood for each one of the panel structures under study. Interval plots for 95% CI for the mean values of the MOR and MOE for all the plywood structures are displayed in Figures 1 and 2, respectively. As it was expected with increasing density the values of MOR and MOE of the plywood samples increased (Figure 3).

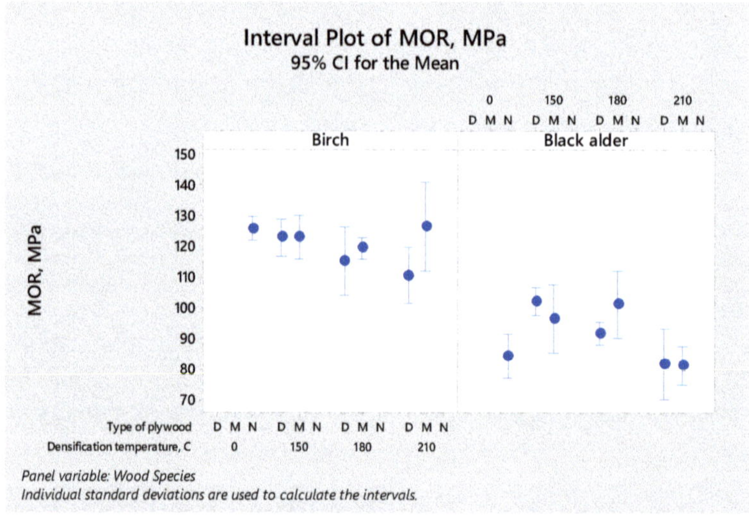

Figure 1. Variation of the bending strength (MOR) of birch and black alder plywood as a function of densification temperature of veneers prior to plywood manufacturing and the type of plywood: Made of densified (D), non-densified (N), and mixed (M) veneers.

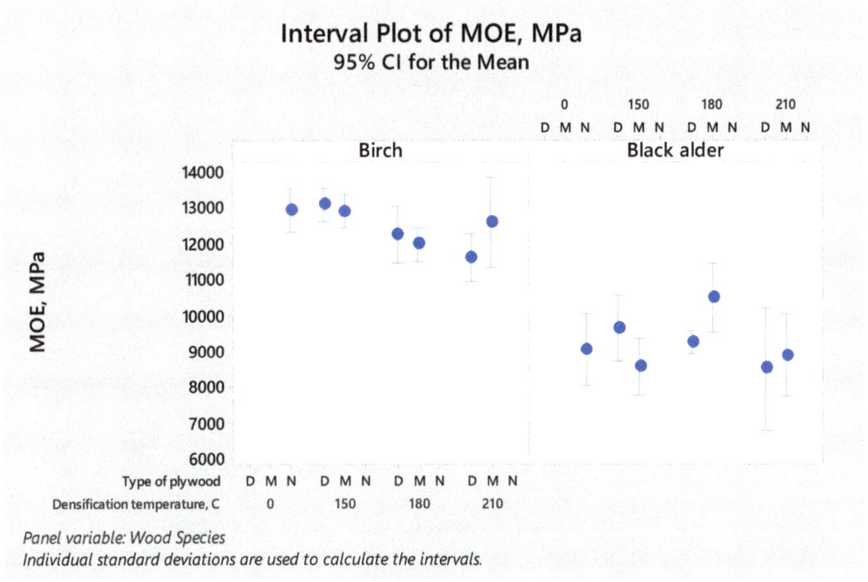

Figure 2. Variation of the modulus of elasticity (MOE) of birch and black alder as a function of densification temperature of veneers prior to plywood manufacturing and the type of plywood: Made of densified (D), non-densified (N), and mixed (M) veneers.

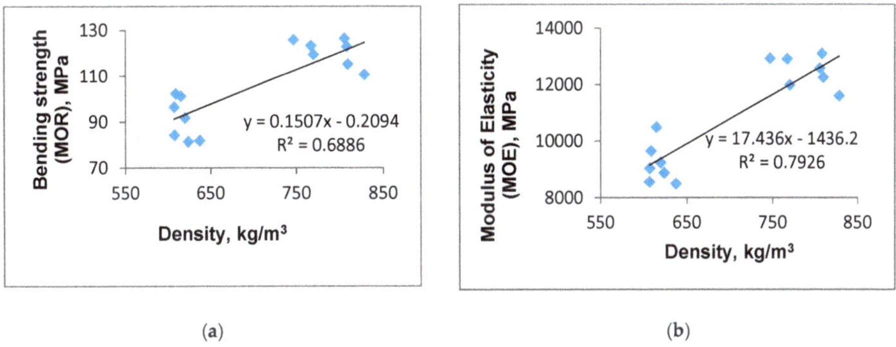

(a)   (b)

Figure 3. Relationship between the bending strength (MOR)—(**a**), modulus of elasticity (MOE)—(**b**) and density of plywood panels.

The birch plywood N had higher average values of MOR and MOE compared to all other average values obtained for the birch panels D made of densified veneers. A peak value for MOE of about 13,105 MPa was determined for the birch plywood at the temperature densification level of 150 °C. It appears that the densification applied to birch veneers before plywood manufacturing did not show any essential enhancement in terms of panel bending properties.

Similar trends were found by Bekhta et al. for beech plywood samples [1]. The MOR values of beech plywood made of non-densified veneers presented higher values by an average of 8.2% for the panels made of thicker veneer and by 16.4% for the panels with thinner veneer when compared to the results for beech plywood made of non-densified veneers. The veneers undergo two times a densification, once during the specific densification process and later in the process of plywood manufacturing, which produced the weakening of bending strength.

The adhesive spread, even almost 50% lower than the amount used in industry, did not show much difference between the bending properties of the birch plywood panels made of densified veneers or combinations of veneers with and without densification.

But in the case of black alder plywood the MOR and MOE values were higher for panels made of densified veneers glued using a low glue spread of 60 g/m^2 than the ones of control non-densified veneers glued with a spread amount of 80 g/m^2.

For black alder plywood the densification process applied to veneers at the temperature levels of 150 °C and 180 °C resulted in increased values with 21.42% and 9.19% for MOR, and 6.75% and 2.28% for MOE, when they are compared to control panel N, respectively.

Results of the samples determined in this study are comparable to those found in previous studies. Bekhta et al. found a MOR value of 123.8 MPa for 5-ply birch plywood made of densified veneers with cold-press using the PF adhesive at a glue spread of 135 g/m^2 [11]. When using the same PF adhesive at a higher spread of 160 g/m^2 for the birch plywood made of densified veneers a MOR value of 161 MPa and a MOE value of 14,883 MPa were found [25]. In the case of black alder plywood made of densified veneers glued with UF adhesive at a spread rate of 160 g/m^2 a higher MOR value of 119.8 MPa when compared to the results of the present study was achieved [26].

The results of bending properties showed that the increase of veneer densification temperature resulted in a gradually decrease of MOR and MOE values for the plywood panels made of densified veneers of both species. The same trend was noticed for almost all mixed panels.

For birch plywood only small differences were noticed between the MOR and MOE values of plywood panels D compared to M for the temperature levels of veneer densification of 150 °C and 180 °C. Thus, a slight and gradual increase of about 0.08%, 3.47%, and 14.54% in MOR values was noticed for panels M compared to panel D at the same temperature level of veneer densification, increasing from 150 °C to 180 °C, and then to 210 °C, respectively. The highest increase of 8.42% for the MOE value of the birch panels M from D was found at the temperature densification level of 210 °C.

For black alder plywood panels D compared to M, the MOR and MOE values subsequently decreased and increased with slow rates for different temperature levels of veneer densification. Thus, given the temperature densification level of 150 °C of black alder veneers, the MOR value for panel M was 5.88% lower than the one for panel D.

But the MOR and MOE values increased with 10.98% and 13.45% from panel D to panel M at the temperature densification level of 180 °C, respectively. No relevant changes were determined for the black alder panels M and D at the veneer densification level of 210 °C.

The highest MOE value of 10,482 MPa was determined for black alder mixed panel M at the veneer densification level of 180 °C.

The strength values are significantly affected by the density of the samples. Overall, very low differences in densities and bending properties have been noticed for plywood panels D and M for both species under study. The outer densified layer is responsible for the bending strength. In a previous study for mixed plywood panels made of densified and non-densified beech veneers a decrease in MOR values when compared to panels made of densified veneers was found [1]. The values for bending strength and modulus of elasticity obtained for all the plywood structures were found higher than the limit values for structural plywood panels.

3.3. Shear Strength of the Plywood Samples

The shear strength mean values of all plywood panels were above the limit of 1 MPa indicated by the EN 314-2 [23]. All plywood panels manufactured in this study presented satisfactory bonding value for indoor applications (Figure 4).

The mean shear strength value of the birch panels made of non-densified veneers N (2.40 MPa) was found lower than that one for black alder control panels (2.98 MPa).

As regards to the other two types of panels, D and M, the shear strength values were found higher for birch plywood than for black alder plywood when considering the same plywood structure type.

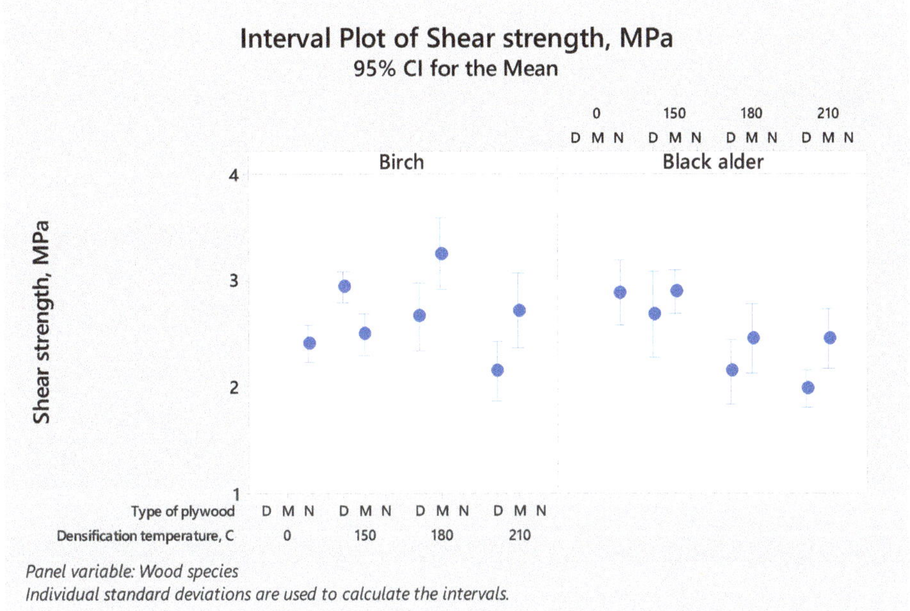

Figure 4. Variation of the shear strength of birch and black alder plywood as a function of densification temperature of veneers prior to plywood manufacturing and the type of plywood: Made of densified (D), non-densified (N), and mixed (M) veneers.

In the case of birch plywood D, the densification of veneers applied prior plywood manufacturing produced an enhancement of the bonding quality with 22.9% and 10.83% for the panels made of densified veneers at the temperature level of 150 °C and 180 °C, while the temperature level of densification at 210 °C did not show a better shear strength compared to birch control panel N. The highest mean value for the shear strength of panels D of about 2.95 MPa was obtained for panels at the densification temperature level of 150 °C.

Such results are in accordance with those found in the specialty literature. The densification of veneers produces a smoother surface that was proved to enhance the shear strength of plywood samples [26]. It was also proved that the adhesive spread for such smooth veneers should be lower than for the non-densified veneers and the final product can meet the standard requirements for plywood panels [1,26]. This statement applies to the present study as well.

In a previous work the shear strength increased from 0.77 to 3.22 MPa for beech plywood produced from densified veneers with a UF glue spread of 110 g/m^2 when compared to plywood of non-densified veneers, while for a higher glue spread of 180 g/m^2, only a small increase of 12.69% was determined [1].

In terms of shear strength, values in the same range with the results in this work for birch plywood, but with higher glue spread rates than the ones used in the current study, can be found in the literature [11,25–27].

In the case of black alder plywood made of densified veneers no improvement in terms of shear strength quality compared to the results for control panels was noticed.

The highest mean value for the shear strength of 2.89 MPa was obtained for panels of non-densified veneers N and mixed panels M at the densification temperature level of 150 °C.

Overall, the shear strength values presented little differences between panels M and D for both species, in most of the cases M was higher than D. An exception was found in the case of birch panels M that had a lower shear strength value (2.49 MPa) compared to that one of panels D (2.95 MPa) at the same densification temperature level of 150 °C.

It appears that the better bonding strength of samples with mixed structure M when compared to panels D is due to the core (n-d-n) that acts like a compact adhesive layer. The higher glue spread of 70 g/m^2 compared to the one in panels D (60 g/m^2) gives a good bonding to the outer densified layers.

The shear strength values gradually decreased with the increase of temperature level for veneer densification.

Increased rates of 25.3% and 26.51% of the shear strength values for birch panels M compared to D at the densification temperature levels of 180 °C and 210 °C, respectively, were found. In the case of black alder gradually increased rates of about 7.83%, 14.48%, and 24.24% of the bonding quality of panels M compared to D at the densification temperature levels of 150 °C, 180 °C, and 210 °C, respectively, were determined.

The highest mean value for the shear strength of mixed birch panels of about 3.26 MPa was found for panels M at the densification temperature level of 180 °C, while for black alder plywood the highest mean value was of about 2.89 MPa for control panels N.

3.4. ANOVA Analysis and Response Optimization

Multifactorial ANOVA analysis (Table 3) was applied to determine in what extent the mechanical properties of plywood panels were influenced by the wood species, densification temperature applied to veneer before plywood manufacturing and the type of plywood produced and pairs of these factors as well. The analysis showed that the wood species and densification temperature along with their interactive cumulative effect could be considered statistically meaningful for the strength properties (p-value ≤ 0.05), while the type of plywood presented less significance. The test statistics showed that the densification temperature was found significant for MOR at 90% CI (p-value ≤ 0.1). The results showed that the wood species is not a statistically significant factor for the shear strength. Such results are in accordance with the statements previously presented and discussed in this work.

Table 3. Multifactorial ANOVA analysis of variance for the mechanical properties (Wood species = A; Densification temperature = B).

One-Way ANOVA Response	Source	Degrees of Freedom (DF)	Adjusted Sums of Squares (Adj SS)	Adjusted Means Squares (Adj MS)	F-Value	*p*-Value
Bending strength (MOR), MPa	A	1	30,315	30,314.7	199.16	0.000
	B	3	2538	846.1	5.56	0.001
	A*B	3	2361	787.0	5.17	0.002
Modulus of elasticity (MOE), MPa	A	1	363,249,244	363,249,244	213.91	0.000
	B	3	11,208,650	3,736,217	2.20	0.091
	A*B	3	16,538,180	5,512,727	3.25	0.024
Shear strength (SS), MPa	A	1	0.427	0.4272	1.05	0.306
	B	3	7.929	2.6431	6.52	0.000
	A*B	3	10.358	3.4526	8.52	0.000

In order to get the best plywood performance the manufacturing parameters need optimization. Based on the results of this study it appears that for both species the densification of veneers applied at 150 °C provides the best results in terms of bending properties and a good bonding quality of the plywood panels type D and M; such properties have been achieved using a low glue consumption of about 60 and 80 g/m^2, respectively. No other temperature level higher than 150 °C is needed for veneers densification prior to plywood manufacturing.

Table 4 presents the response optimization when selected between species to obtain the best values predicted for the bending properties (MOR and MOE) and the shear strength of the plywood panels. The result shows that the mixed birch panels made of non-densified and densified veneers at the temperature level of 154.28 °C were found to have the best values of their bending and bonding properties.

Table 4. Response optimization between species for the maximum strength values of plywood samples.

Solution	Densification Temperature, °C	Wood Species	Type of Plywood	Properties		Composite Desirability
1	154.286	Birch	M	MOE, Mpa Fit	12,830.8	0.765479
				MOR, Mpa Fit	127.384	
				Shear strength, Mpa Fit	2.97644	0.569985

4. Conclusions

In this study the effect of veneer densification temperature and wood species on the plywood properties when manufactured from non-densified, densified and from alternate layers of densified and non-densified veneers has been analyzed.

- The bending strength and the modulus of elasticity along with overall shear strength presented higher values for birch plywood compared to black alder plywood. They increased with the increasing in density.
- The densification applied to birch veneers compared to black alder presented little improvement in terms of panel bending properties but quite significant enhancement of the bonding quality. The densification of black alder veneer did not provide any improvement in terms of shear strength quality of the plywood panels.
- The increase of veneer densification temperature resulted in a gradually decrease of MOR, MOE, and shear strength values for the plywood panels made of densified veneers and mixed panels of both species. The temperature of about 150 °C for veneer densification seemed to be enough to provide the best values for bending and bonding properties of such panels.
- All plywood panels in this study were manufactured using reduced glue consumption and they presented satisfactory properties performance for indoor applications. Apart of the plywood properties performance, such an approach can provide low emissions of toxic compounds and low costs for the finished product.
- In further studies it would be desirable to evaluate some of the other properties of such samples including thickness swelling and water absorption would help better understanding behavior of these panels. Also some data on surface quality evaluation of non-densified and densified veneer sheets could help to optimize overall adhesive consumption.

Author Contributions: Conceptualization, methodology, software, validation, formal analysis, P.B., E.-A.S. and Y.S.; investigation, resources, and data curation, P.B. and E.-A.S.; writing—original draft preparation, E.-A.S.; writing—review and editing, E.-A.S. and P.B.; visualization and supervision, P.B.; supervision, P.B. All authors have read and agreed to the published version of the manuscript.

Funding: Part of this research was funded by the COST Action FP1407/STSM 41745.

Acknowledgments: We would like to thank for the support offered by the Laboratory of Wood-Based Composites, Cellulose & Paper from the Ukrainian National Forestry University to perform this research.

Conflicts of Interest: The authors declare no conflict of interest.

References

1. Bekhta, P.; Salca, E.A.; Lunguleasa, A. Some properties of plywood panels manufactured from combinations of thermally densified and non-densified veneers of different thickness in one structure. *J. Build. Eng.* **2020**, *29*, 101116. [CrossRef]
2. Daoui, A.; Descamps, C.; Marchal, R.; Zerizer, A. Influence of veneer quality on beech LVL mechanical properties. *Maderas Cienc Tecnol.* **2011**, *13*, 69–83. [CrossRef]
3. Del Menezzi, C.; Mendes, L.; de Souza, M.; Bortoletto, G. Effect of Nondestructive Evaluation of Veneers on the Properties of Laminated Veneer Lumber (LVL) from a Tropical Species. *Forests* **2013**, *4*, 270–278. [CrossRef]

4. de Melo, R.R.; Del Menezzi, C.H.S. Influence of veneer thickness on the properties of LVL from Parica (*Schizolobium amazonicum*) plantation trees. *Eur. J. Wood Wood Prod.* **2014**, *72*, 191–198. [CrossRef]

5. Pałubicki, B.; Marchal, R.; Butaud, J.-C.; Denaud, L.-E.; Bléron, L.; Collet, R.; Kowaluk, G. A method of lathe checks measurement; SMOF device and its software. *Eur. J. Wood Wood Prod.* **2010**, *68*, 151–159. [CrossRef]

6. DeVallance, D.B.; Funck, J.W.; Reeb, J.E. Douglas-Fir plywood gluebond quality as influenced by veneer roughness, lathe checks, and annual ring characteristics. *Forest Prod. J.* **2007**, *57*, 21–28.

7. Wang, B.J.; Ellis, S.; Dai, C. Veneer surface roughness and compressibility pertaining to plywood/LVL manufacturing. Part II. Optimum panel densification. *Wood Fiber Sci.* **2006**, *38*, 727–735.

8. Seborg, R.M.; Millet, M.; Stamm, A.J. Heat-stabilized compressed wood. *Staypack. Mech. Eng.* **1945**, *67*, 25–31.

9. Morsing, N. Densification of Wood. The Influence of Hygrothermal Treatment on Compression of Beech Perpendicular to the Grain. Ph.D. Thesis, Building Materials Laboratory, Technical University of Denmark, Lyngby, Denmark, 1997.

10. Navi, P.; Sandberg, D. *Thermo-Hydro-Mechanical Wood Processing. Lausanne*; EPFL Press: Lausanne, Switzerland, 2012.

11. Bekhta, P.; Hiziroglu, S.; Shepelyuk, O. Properties of plywood manufactured from compressed veneer as building material. *Mater. Des.* **2009**, *30*, 947–953. [CrossRef]

12. Kutnar, A.; Kamke, F.A.; Sernek, M. Density profile and morphology of viscoelastic thermal compressed wood. *Wood Sci. Technol.* **2009**, *43*, 57–68. [CrossRef]

13. Arruda, L.M.; Menezzi, H.S.C. Properties of a laminated wood composite produced with thermomechanically treated veneers. *Adv. Mater. Sci. Eng.* **2016**, 9. [CrossRef]

14. Bekhta, P.; Niemz, P.; Sedliacik, J. Effect of pre-pressing of veneer on the glueability and properties of veneer-based products. *Eur. J. Wood Wood Prod.* **2012**, *70*, 99–106. [CrossRef]

15. Bekhta, P.; Salca, E.A. Influence of veneer densification on the shear strength and temperature behavior inside the plywood during hot press. *Constr. Build. Mater.* **2018**, *162*, 20–26. [CrossRef]

16. Bekhta, P.; Sedliacik, J.; Jones, D. Effect of short-term thermomechanical densification of wood veneers on the properties of birch plywood. *Eur. J. Wood Wood Prod.* **2018**, *76*, 549–562. [CrossRef]

17. Candan, Z.; Hiziroglu, S.; Mcdonald, A.G. Surface quality of thermally compressed Douglas fir veneer. *Mater. Des.* **2010**, *31*, 3574–3577. [CrossRef]

18. Diouf, P.N.; Stevanovic, T.; Cloutier, A.; Fang, C.-H.; Blanchet, P.; Koubaa, A.; Mariotti, N. Effects of thermo-hygro-mechanical densification on the surface characteristics of trembling aspen and hybrid poplar wood veneers. *Appl. Surf. Sci.* **2011**, *257*, 3558–3564. [CrossRef]

19. Li, W.; Wang, C.; Zhang, Y.; Jia, C.; Gao, C.; Jin, J. The influence of hot compression on the surface characteristics of poplar veneer. *BioRes* **2014**, *9*, 2808–2823. [CrossRef]

20. EN 323. *Wood-Based Panels—Determination of Density*; European Committee for Standardization: Brussels, Belgium, 1993.

21. EN 310. *Wood-Based Panels—Determination of Modulus of Elasticity in Bending and of Bending Strength*; European Committee for Standartiztion: Brussels, Belgium, 1993.

22. EN 314-1. *Plywood—Bonding Quality—Part 1: Test Methods*; European Committee for Standardization: Brussels, Belgium, 2004.

23. EN 314-2. *Plywood—Bonding Quality—Part 2: Requirements*; European Committee for Standardization: Brussels, Belgium, 1993.

24. EN 315. *Plywood. Tolerances for Dimensions*; European Committee for Standardization: Brussels, Belgium, 2000.

25. Bekhta, P.; Sedliacik, J.; Bekhta, N. Effects of selected parameters on the bonding quality and temperature evolution inside plywood during pressing. *Polymers* **2020**, *12*, 1035. [CrossRef] [PubMed]

26. Bekhta, P.; Sedliacik, J. Environmetally-friendly high-density polyethylene-bonded plywood panels. *Polymers* **2019**, *11*, 1166. [CrossRef] [PubMed]

27. Bekhta, P.; Marutzky, R. Reduction of glue consumption in the plywood production by using previously compressed veneer. *Holz Roh Werkst* **2007**, *65*, 87–88. [CrossRef]

 *forests*

Article

Development of Wood Composites from Recycled Fibres Bonded with Magnesium Lignosulfonate

Petar Antov, George I. Mantanis and Viktor Savov

[1] Department of Mechanical Wood Technology, Faculty of Forest Industry, University of Forestry, 1797 Sofia, Bulgaria; victor_savov@ltu.bg

[2] Department of Forestry, Wood Sciences and Design, Lab of Wood Science and Technology, University of Thessaly, 43100 Karditsa, Greece; mantanis@uth.gr

* Correspondence: p.antov@ltu.bg

Received: 13 May 2020; Accepted: 25 May 2020; Published: 1 June 2020

Abstract: The potential of producing ecofriendly composites from industrial waste fibres, bonded with magnesium lignosulfonate, a lignin-based formaldehyde-free adhesive, was investigated in this work. Composites were produced in the laboratory using the following parameters: a hot press temperature of 210 °C, a pressing time of 16 min, and a 15% gluing content of magnesium lignosulfonate (on the dry fibres). The physical and mechanical properties of the produced composites were evaluated and compared with the European Standard (EN) required properties (EN 312, EN 622-5) of common wood-based panels, such as particleboards for internal use in dry conditions (type P2), load-bearing particleboards for use in humid conditions (type P5), heavy-duty load-bearing particleboards for use in humid conditions (type P7), and medium-density fibreboards (MDF) for use in dry conditions. In general, the new produced composites exhibited satisfactory mechanical properties: a bending strength (MOR) (18.5 N·mm^{-2}) that was 42% higher than that required for type P2 particleboards (13 N·mm^{-2}) and 16% higher than that required for type P5 particleboards (16 N·mm^{-2}). Additionally, the modulus of elasticity (MOE) of composites (2225 N·mm^{-2}) was 24% higher than that required for type P2 particleboards (1800 N·mm^{-2}) and equivalent to the required MOE of MDF panels for use in dry conditions (2200 N·mm^{-2}). However, these ecofriendly composites showed deteriorated moisture properties, i.e., 24 h swelling and 24 h water absorption, which were a distinct disadvantage. This should be further investigated, as modifications in the lignosulfonate formula used and/or production parameters are necessary.

Keywords: wood composites; waste fibres; magnesium lignosulfonate; bioadhesives

1. Introduction

Optimisation of resource efficiency is one of the key objectives to implement the principles of circular economy and face the challenges of increased demand for wood and wood-based products worldwide. Cascading use of lignocellulosic resources, defined as "the efficient utilisation of resources by using residues and recycled materials for material use to extend total biomass availability within a given *system*", is one of the leading principles for achieving this goal [1,2]. Every year, pulp and paper industries generate significant quantities of nonhazardous solid waste and sludge, which require efficient utilisation as waste materials or by-products [3,4]. Pulp and paper sludge contains fibres and can therefore be re-used for the manufacturing of new composites [5–9].

Conventional adhesive systems used for wood composites are commonly made of fossil-derived constituents, which are based on formaldehyde, urea, phenol, melamine, and/or isocyanates [10–13]. Due to their high reactivity, chemical versatility, cost effectiveness, and technological performance regarding their strength of adhesion and moisture resistance, formaldehyde-based polycondensation adhesives have been widely used in wood-based panel industries [14–20]. Nonetheless, they have a major disadvantage, that is, their hazardous emission of volatile organic compounds (VOCs), such as formaldehyde, from finished panel products, especially in indoor environments [21,22]. In 2004, formaldehyde was reclassified from "probable human carcinogen" to "known human carcinogen" by the International Agency for Research on Cancer [23]. This, combined with increased environmental concerns, has changed the trend in the global wood industry from synthetic formaldehyde-based adhesives towards the development of biobased adhesives for the production of green composites [11–13,24–35]. This term, "green composites", refers to composites fabricated from both natural fibres and a biobased matrix [36,37]. Thus, such new wood composites, having acceptable physical and mechanical properties, can offer several important advantages, such as renewability, biodegradability, and lower production costs [26,32,33,38–43].

Following cellulose, lignin is the second most abundant polymer encountered in nature [44]. With an estimated 300 billion tons in the biosphere and an annual re-synthesis of ca. 20 billion tons [45,46], it is present in wood, grass, agricultural residues, and other plants [47]. In addition, lignin is considered as a valuable byproduct of the chemical pulping process of wood (i.e., lignosulfonates originating from the sulphite pulping process and kraft lignin from the sulphate pulping process), with an annual accumulation of more than 50–75 million tons worldwide [27,48,49]. Nevertheless, only 10% of technical lignin is further utilised industrially; the rest of it is mainly incinerated for energy purposes or remains unutilised [50–52]. Thus, lignin appears to be an abundant and renewable raw material for value-added chemicals [53], as well as for wood adhesives [12,13,26,32,54–61]. Due to its phenolic structure, lignin has favourable properties, such as high hydrophobicity and low polydispersity, for the formulation of new adhesives suitable for the wood-based panel industry [62]. However, the aromatic structure of lignin lowers the reactivity of the resin, which represents a significant drawback in applications where fast-curing times are needed.

Lignosulfonates (R-SO3H), the salts of lignin sulfonic acid, are one of the main sources of technical lignins. They are water-soluble anionic polyelectrolytes obtained as the byproducts of the sulphite process in which delignification of wood is performed by means of HSO_3^- and SO_3^{2-} ions [63,64]. Lignosulfonates derived from sulphite lignin processing have a relatively high molecular weight (15,000–50,000 g·mol^{-1}), a high content of inorganics (up to 25%), and a sulphur content of approx. 8% [49,65]. Lignosulfonates are produced as dry solids in relatively large quantities, e.g., the annual production is approximately 1.1 million tons [66,67]. Increased scientific and industrial interest in lignin-originating compounds, including lignosulfonates, is due to the phenolic structure of lignin allowing partial replacement of phenol (C_6H_6O) in the formulation of lignin–phenol–formaldehyde (LPF) resins to formulate adhesives for wood products [12,28,68–71], in order to decrease production costs and reduce toxicity. The use of unpurified sulphur-containing lignosulfonates results in lower reactivity of the resin, leading to extended pressing times [72,73]. In order to use lignosulfonates as main components of adhesives, this lower reactivity should be compensated by a suitable cross-linker [26] or by modifying the parameters for the production of new composites [74,75].

The aim of this research work was to investigate the potential of producing ecofriendly composites from waste wood fibres, bonded with a lignin-based adhesive, namely, magnesium lignosulfonate, complying with the requirements of the European Standards.

2. Materials and Methods

Industrial waste fibre mass (Figure 1) composed of two softwood species, namely, Scots pine (*Pinus silvestris* L.) and Norway spruce (*Picea abies* Karst.), oven dried to 12% moisture content, was supplied by the company Mondi Stambolyiski EAD, a Bulgarian paper mill (Stambolyiski, Bulgaria). The bulk density of it was 4468 kg·m^{-3}. The main fibre fraction appeared to have a length between 500 and 1000 μm and a reduced lignin content of 7%.

Figure 1. Industrial waste fibre mass (photo by V. Savov).

Magnesium lignosulfonate additive (CAS No. 8061-54-9), which is named Borresperse 390 Paper (Borregaard, Germany), was utilised as a main binder at 15% gluing content on the dry weight of the fibres. This additive had the following characteristics: magnesium content: 6%; reduced sugars: 7%; sulphate content: 2%; and total solids content: 51.2%. European beech (*Fagus sylvatica* L.) veneer sheets, having a thickness of 1.1 mm and a moisture content of approximately 8%, provided by the company Welde Bulgaria AD (Troyan, Bulgaria), were used for veneering the surface layers of the composites.

Firstly, fibreboards were produced in the laboratory at a thickness of 16 mm and a target density of 720 kg·m^{-3}. Industrial waste fibres were mixed with the lignosulfonate-based adhesive in a high-speed laboratory glue blender (850 min^{-1}). Hot pressing was performed in a laboratory press (PMC ST 100, Italy). The press temperature used was 210 °C. The pressing regime applied consisted of the following four stages: In the first stage, the pressure was increased to 4.5 MPa for 1 min, then, it was gradually decreased to 2.23 MPa for 3 min, followed by decreasing the pressure to 0.74 MPa for 10 min. The last pressing period was implemented at a pressure of 1.78 MPa for 2 min. Following hot pressing, the produced fibreboards were conditioned in a laboratory acclimatising chamber at 20 °C and 60% relative humidity, for a period of 7 days.

After that, the fibreboards were veneered with beech veneer sheets, using magnesium lignosulfonate as a binder, at a gluing content of 80 g/m^2. The applied veneering press factor was 1 min mm^{-1} of panel thickness, at a pressure of 0.6 MPa and a press temperature of 200 °C.

The physical and mechanical properties of the new composites were determined according to the European Standards EN 310, EN 317, EN 322, and EN 323 [76–79]. Thickness swelling and water absorption tests were carried out for 24 h. Precision balance (Kern, Germany) with a readability of 0.01 g and a digital calliper with a readability of 0.01 g were used to determine the mass and dimensions of the test specimens, respectively. For measurement of the mechanical properties of the composites, a universal-material testing machine Zwick/Roell Z010 was used.

Formaldehyde emission of the fabricated composites was measured in the laboratory of the factory Kronospan Bulgaria EOOD (Veliko Tarnovo, Bulgaria) on four samples using the perforator method [80].

In the last part, variational and statistical analysis of the results was performed with the specialised software QstatLab 6.0.

3. Results and Discussion

3.1. Formaldehyde Emission

The formaldehyde emission of the produced composites was estimated to be 1.1 mg/100 g (±0.1), according to the standard EN ISO (International Organization for Standardization) 12460-5 (2015). This emission value is exceptionally low and is considered as a zero formaldehyde value [13,80]. Formaldehyde is a naturally occurring compound in wood, formed by its main polymeric components (i.e., cellulose, hemicellulose, and lignin) or extractives [81,82]. Thus, the determined formaldehyde emission value, which is much lower than 2 mg/100 g, i.e., the emission of natural wood [13,26], allows for the defining of fabricated composites as ecofriendly composites.

3.2. Physical and Mechanical Properties

A summary of the physical and mechanical properties of the composites, comprising recycled wood fibres and magnesium lignosulfonate, is shown in Table 1. The respective variational and statistical data are also presented. The density of the composites varied from 681 to 772 kg·m^{-3}, relatively close to the projected value. The difference in this main characteristic of the composites was significantly below 5%; thus, this will not have an effect on the other physical and mechanical properties.

Table 1. Physical and mechanical properties of the laboratory-produced composites.

Property	Average (Mean Value) $\overline{X}$	Standard Deviation S_x	Standard Error m_x	Probability (p-Value) P_x, %	Sample Size
Density ρ, kg·m^{-3}	743	23.9	7.5	1.0	10
Water absorption (24 h) A, %	168.4	12.9	4.1	2.4	10
Thickness swelling (24 h) Gt, %	82.8	5.9	1.9	2.3	10
Bending strength (MOR) fm, N·mm^{-2}	18.5	1.6	0.6	3.1	8
Modulus of elasticity (MOE) Em, N·mm^{-2}	2255	228	81	3.6	8

In order to analyse the technical properties of the produced composites, which were fabricated from recycled lignocellulosic fibres, bonded with magnesium lignosulfonate, and veneered with beech veneers (experimental boards; labelled as Eb), their physical and mechanical properties were compared with the minimum required properties of common wood-based panels, such as (i) particleboards for internal use (including furniture) in dry conditions (type P2), (ii) load-bearing particleboards for use in humid conditions (type P5), (iii) heavy-duty load-bearing particleboards for use in humid conditions (type P7) [83], and (iv) medium-density fibreboards (MDF) for use in dry conditions [84]. Particleboards of type P2 and MDF were selected, since they are the most commonly used panels. Particleboards of types P5 and P7 were selected because of the very stringent requirements in respect of their mechanical and swelling properties.

As is known in the art [10,11,13], water absorption (WA) and thickness swelling (TS) are critical physical properties, strongly related with the dimensional stability of wood-based composites. WA is not a standardised technical property; nonetheless, according to the literature [85], the WA of common MDF typically varies between 50% and 70%. In this work, the WA of the laboratory-fabricated composites was found to be approximately 168%, i.e., three times higher than that of a standard-grade MDF panel. Consequently, magnesium lignosulfonate, as a lignin-based compound, will require a chemical modification, among others, in order to increase its chemical reactivity and bonding efficiency [49,70].

A graphical representation of the TS of the compared wood-based composites is presented in Figure 2. The particleboard for internal use (type P2) is not included, as the thickness swelling limit is not required by the standard EN 312 (2010).

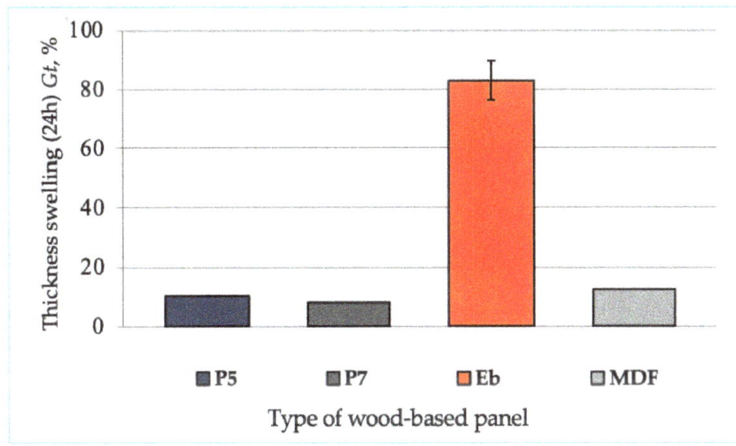

Figure 2. Thickness swelling of composites produced (experimental board: Eb). (Error bar represents the standard deviation.)

As seen in Figure 2, the composite Eb exhibited very high swelling values, i.e., 6 times higher than the minimum limit of MDF for use in dry conditions, 7 times higher than the minimum limit of load-bearing particleboards for use in humid conditions (type P5), and 9 times higher than the minimum limit of heavy-duty load-bearing particleboards for use in humid conditions (type P7).

In addition, a graphical representation of the mean bending strength (MOR) of composite Eb and the minimum required MOR limit values of typical wood-based panels is shown in Figure 3.

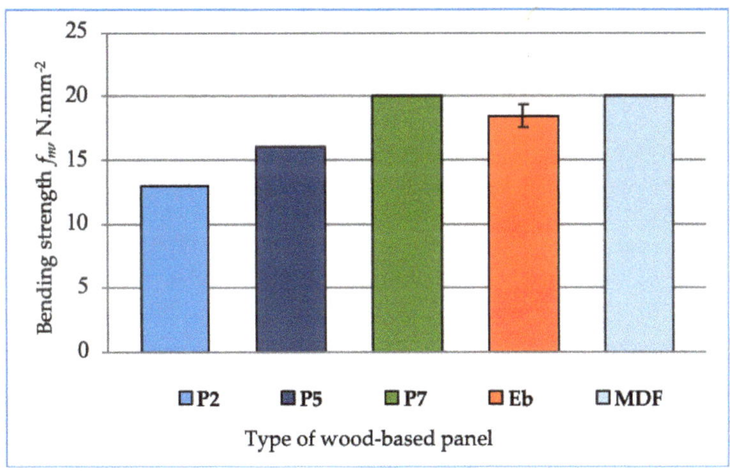

Figure 3. Bending strength (MOR) of the produced composite panels vs. MOR limit values of typical wood-based panels. (Error bar represents the standard deviation.)

The produced ecofriendly composite Eb exhibited very satisfactory bending strength properties, i.e., the MOR was 42% higher than the minimum required MOR value of particleboards for interior fitments (type P2; 13 N·mm^{-2}). Composite Eb showed higher MOR values as compared with the minimum accepted MOR for load-bearing particleboards for use in humid conditions (type P5; 16 N·mm^{-2}). Furthermore, ecofriendly composite Eb exhibited a mean MOR strength of 18.5 N·mm^{-2}, which was slightly lower than the minimum MOR limit for MDF panels (i.e., 20 N·mm^{-2}) and the minimum MOR for heavy-duty load-bearing particleboards for use in humid conditions (type P7; 20 N·mm^{-2}). Thus, these MOR values were quite comparable. It should be noted that, typically, P7-type particleboards are heavy or very heavy wood panels, while the composite Eb produced in this work had significantly lower densities.

Finally, a graphical representation of the average modulus of elasticity (MOE) of composite Eb and the minimum required MOE limit values of selected wood-based panels is shown in Figure 4.

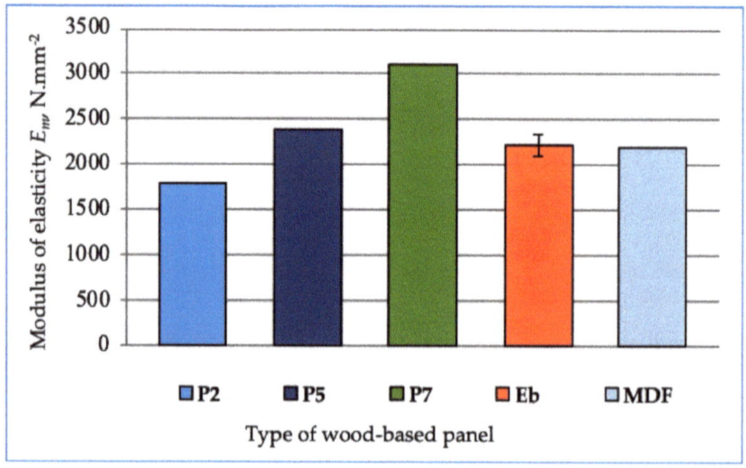

Figure 4. Modulus of elasticity (MOE) of the produced composite panels vs. MOE limit values of typical wood-based panels. (Error bar represents the standard deviation.)

The mean MOE of the composite Eb was found to be ca. 2225 N·mm^{-2}. This value is 24% larger than the minimum accepted for particleboards for interior fitments for use in dry conditions (type P2; 1800 N·mm^{-2}) and almost equivalent to the MOE limit value of MDF for use in dry conditions (2200 N·mm^{-2}). The composite Eb showed a lower MOE property as compared with the minimum MOE limit of load-bearing particleboards for use in humid conditions (type P5; 2400 N·mm^{-2}) and heavy-duty load-bearing particleboards for use in humid conditions (type P7; 3100 N·mm^{-2}). Laboratory-made composites cannot be utilised for load-bearing applications, in which quite an increased load resistance is needed. As a whole, it is a positive outcome that developed ecofriendly composites exhibited much higher strength properties than the minimum required for particleboards for interior fitments, including furniture (type P2), and similar strength properties as compared to common MDF boards.

4. Conclusions

Ecofriendly composites with acceptable physical and mechanical properties according to ENs, except for thickness swelling, may be produced from industrial waste fibres, bonded with a lignin-based formaldehyde-free adhesive, namely, magnesium lignosulfonate. The composites were manufactured in the laboratory using 15% gluing content of magnesium lignosulfonate as a binder, following the covering of composites with beech veneers, also glued with magnesium lignosulfonate. As above, laboratory-fabricated composites showed quite satisfactory mechanical properties, such as MOR and MOE, which were higher than the minimum required for particleboards for interior fitments for use in dry conditions (type P2) and equivalent to the least required for MDF panels of standard grade.

Markedly, the produced ecofriendly composites had "almost zero" emission of formaldehyde (ca. 1.1 mg per 100 g of composite), as measured by the standard perforator method.

Nonetheless, the produced composites exhibited a deteriorated dimensional stability, as found in panel properties like thickness swelling and water absorption (24 h), which was a major drawback; this disadvantage could be resolved by modifications of the lignosulfonate formula. Therefore, future research should focus on optimising the production and pressing parameters, as well as further investigating the bonding mechanism between lignosulfonate and fibres.

Nevertheless, new ecofriendly composites represent promising alternatives to conventional wood-based panels, which utilise only petroleum-derived adhesives. As this research showed, waste fibres from the pulp and paper industries, as raw materials, can be efficiently utilised, and industrial waste discharge can be reduced, thus contributing to the implementation of circular economy principles.

Author Contributions: Conceptualisation, P.A. and V.S.; methodology, P.A. and V.S.; validation, P.A. and G.I.M.; investigation, P.A., G.I.M., and V.S.; resources, P.A. and V.S.; writing the original draft preparation, P.A. and G.I.M.; Writing—review and editing, G.I.M. and P.A.; visualisation, P.A. and V.S.; supervision, P.A. and G.I.M.; project administration, P.A. All authors have read and agreed to the published version of the manuscript.

Funding: This research work was supported by project no. НИС-Б-1002/03.2019 *"Exploitation Properties and Possibilities for Utilisation of Eco-friendly Bio-composite Materials"*, which was carried out at the facilities of the University of Forestry (Sofia, Bulgaria), in cooperation with the University of Thessaly (Department of Forestry, Wood Sciences and Design, Karditsa, Greece).

Conflicts of Interest: The authors declare no conflict of interest.

References

1. Stevulova, N.; Schwarzova, I.; Hospodarova, V.; Junak, J.; Briancin, J. Recycled cellulosic fibres and lignocellulosic aggregates from sustainable building materials. *IJCEE* **2016**, *10*, 716–721.
2. Vis, M.; Mantau, U.; Allen, B. (Eds.) *Study on the Optimised Cascading Use of Wood, No 394/PP/ENT/RCH/14/7689, Final Report*; Publications Office: Brussels, Belgium, 2016; p. 337.
3. Ochoa de Alda, J.A.G. Feasibility of recycling pulp and pulp and paper sludge in the paper and board industries. *Resour. Conserv. Recycl.* **2008**, *52*, 965–972. [CrossRef]
4. Bajpai, P. Generation of Waste in Pulp and Paper Mills. In *Management of Pulp and Paper Mill Waste*; Springer: Berlin/Heidelberg, Germany, 2015; ISBN 978-3-319-11788-1.
5. Davis, E.; Shaler, S.M.; Goodell, B. The incorporation of paper deinking sludge into fibreboard. *For. Prod. J.* **2003**, *53*, 46–54.
6. Geng, X.; Deng, J.; Zhang, S.Y. Pulp and paper sludge as a component of wood adhesive formulation. *Holzforschung* **2007**, *61*, 688–692. [CrossRef]
7. Geng, X.; Deng, J.; Zhang, S.Y. Characteristics of pulp and paper sludge and its utilization for the manufacture of medium density fibreboard. *Wood Fiber Sci.* **2007**, *39*, 345–351.
8. Migneault, S.; Koubaa, A.; Nadji, H.; Riedl, B.; Zhang, S.Y.; Deng, J. Medium-density fibreboard produced using pulp and paper sludge from different pulping processes. *Wood Fiber Sci.* **2010**, *42*, 292–303.
9. Antov, P.; Savov, V. Possibilities for manufacturing eco-friendly medium density fibreboards from recycled fibres–A review. In Proceedings of the 30th International Conference on Wood Science and Technology-ICWST 2019 "Implementation of wood science in woodworking sector" and 70th anniversary of Drvna industrija Journal, Zagreb, Croatia, 12–13 December 2019; ISBN 978-953-292-062-8.
10. Youngquist, J.A. Wood-based composites and panel products. In *Wood Handbook: Wood as an Engineering Material*; USDA Forest Service, Forest Products Laboratory: Madison, WI, USA, 1999; pp. 1–31.
11. Frihart, C.R. Wood adhesion and adhesives. In *Handbook of Wood Chemistry and Wood Composites*; Rowell, R.M., Ed.; CRC Press: Boca Raton, FL, USA, 2005; pp. 214–278.
12. Ferdosian, F.; Pan, Z.; Gao, G.; Zhao, B. Bio-based adhesives and evaluation for wood composites application. *Polymers* **2017**, *9*, 70. [CrossRef]
13. Mantanis, G.I.; Athanassiadou, E.T.; Barbu, M.C.; Wijnendaele, K. Adhesive systems used in the European particleboard, MDF and OSB industries. *Wood Mater. Sci. Eng.* **2018**, *13*, 104–116. [CrossRef]
14. He, G.; Yan, N. Effect of moisture content on curing kinetics of pMDI resin and wood mixtures. *Int. J. Adhes. Adhes.* **2005**, *25*, 450–455. [CrossRef]

15. Jin, Y.; Cheng, X.; Zheng, Z. Preparation and characterization of phenol-formaldehyde adhesives modified with enzymatic hydrolysis lignin. *Bioresour. Technol.* **2010**, *101*, 2046–2048. [CrossRef]

16. Zhang, W.; Ma, Y.; Wang, C.; Li, S.; Zhang, M.; Chu, F. Preparation and properties of lignin-phenol-formaldehyde resins based on different biorefinery residues of agricultural biomass. *Ind. Crops Prod.* **2013**, *43*, 326–333. [CrossRef]

17. Jivkov, V.; Simeonova, R.; Marinova, A. Influence of the veneer quality and load direction on the strength properties of beech plywood as structural material for furniture. *Innov. Woodwork. Ind. Eng. Des.* **2013**, *2*, 86–92.

18. Jivkov, V.; Simeonova, R.; Marinova, A.; Gradeva, G. Study on the gluing abilities of solid surface composites with different wood based materials and foamed PVC. In Proceedings of the 24th International Scientific Conference Wood is Good–User Oriented Material, Technology and Design, Zagreb, Croatia, 18 October 2013; pp. 49–55, ISBN 978-953-292-031-4.

19. Yang, S.; Zhang, Y.; Yuan, T.Q.; Sun, R.C. Lignin-phenol-formaldehyde resin adhesives prepared with biorefinery technical lignins. *J. Appl. Polym. Sci.* **2015**, *132*, 42493. [CrossRef]

20. Solt, P.; Konnerth, J.; Gindl-Altmutter, W.; Kantner, W.; Moser, J.; Mitter, R.; Van Herwijnen, H. Technological performance of formaldehyde-free adhesive alternatives for particleboard industry. *Int. J. Adhes. Adhes.* **2019**, *94*, 99–131. [CrossRef]

21. Kelly, T.J. *Determination of Formaldehyde and Toluene Diisocyanate Emissions from Indoor Residential Sources (No. 97-9)*; California Environmental Protection Agency: Sacramento, CA, USA, 1997.

22. U.S. Consumer Product Safety Commission. *An Update on Formaldehyde (Publication 725)*; U.S. Consumer Product Safety Commission: Bethesda, MD, USA, 2013.

23. International Agency for Research on Cancer. *IARC Classifies Formaldehyde as Carcinogenic to Humans*; IARC: Lyon, France, 2004.

24. Dunky, M. Adhesives based on formaldehyde condensation resins. *Macromol. Symp.* **2004**, *217*, 417–429. [CrossRef]

25. Widyorini, R.; Xu, J.; Umemura, K.; Kawai, S. Manufacture and properties of binderless particleboard from bagasse I: Effects of raw material type, storage methods, and manufacturing process. *J. Wood Sci.* **2005**, *51*, 648–654. [CrossRef]

26. Pizzi, A. Recent developments in eco-efficient bio-based adhesives for wood bonding: Opportunities and issues. *J. Adhes. Sci. Technol.* **2006**, *20*, 829–846. [CrossRef]

27. Kües, U. *Wood Production, Wood technology, and Biotechnological Impacts*; Universitätsverlag Göttingen: Göttingen, Germany, 2007. Available online: https://univerlag.uni-goettingen.de/handle/3/isbn-978-3-940344-11-3 (accessed on 28 February 2020).

28. Papadopoulou, E. Adhesives from renewable resources for binding wood-based panels. *J. Environ. Prot. Ecol.* **2009**, *10*, 1128–1136.

29. Navarrete, P.; Mansouri, H.R.; Pizzi, A.; Tapin-Lingua, S.; Benjelloun-Mlayah, B.; Pasch, H. Wood panel adhesives from low molecular mass lignin and tannin without synthetic resins. *J. Adhes. Sci. Technol.* **2010**, *24*, 1597–1610. [CrossRef]

30. Carvalho, L.; Magalhães, F.; João, F. Formaldehyde emissions from wood-based panels-Testing methods and industrial perspectives. In *Formaldehyde: Chemistry, Applications and Role in Polymerization*; Chan, B.C., Feng, H.L., Eds.; Nova Science Publishers, Inc.: Hauppauge, NY, USA, 2012; pp. 1–45.

31. Vnučec, D.; Kutnar, A.; Goršek, A. Soy-based adhesives for wood-bonding–A review. *J. Adhes. Sci. Technol.* **2016**, *31*, 910–931. [CrossRef]

32. Hemmilä, V.; Adamopoulos, S.; Karlsson, O.; Kumar, A. Development of sustainable bio-adhesives for engineered wood panels-A review. *RSC Adv.* **2017**, *7*, 38604–38630. [CrossRef]

33. Nordström, E.; Demircan, D.; Fogelström, L.; Khabbaz, F.; Malmström, E. Green binders for wood adhesives. In *Applied Adhesive Bonding in Science and Technology*; Interhopen Books: London, UK, 2017; pp. 47–71. [CrossRef]

34. Valyova, M.; Ivanova, Y.; Koynov, D. Investigation of free formaldehyde quantity in production of plywood with modified urea-formaldehyde resin. *Int. J. Wood Des. Technol.* **2017**, *6*, 72–76.

35. Taghiyari, H.R.; Tajvidi, M.; Taghiyari, R.; Mantanis, G.I.; Esmailpour, A.; Hosseinpourpia, R. Nanotechnology for wood quality improvement and protection. In *Nanomaterials for Agriculture and Forestry Applications*; Husen, A., Jawaid, M., Eds.; Elsevier: Amsterdam, The Netherlands, 2020; pp. 469–489.

36. Pfister, D.P.; Larock, R.C. Green composites using switchgrass as a reinforcement for a conjugated linseed oil-based resin. *J. Appl. Polym. Sci.* **2013**, *127*, 1921–1928. [CrossRef]

37. Yuan, Y.; Guo, M.; Liu, F. Preparation and evaluation of green composites using modified ammonium lignosulfonate and polyethylenimine as a binder. *BioResources* **2014**, *9*, 836–848. [CrossRef]

38. El Mansouri, N.E.; Pizzi, A.; Salvado, J. Lignin-based wood panel adhesives without formaldehyde. *Holz. Roh. Werkst.* **2007**, *65*, 65–70. [CrossRef]

39. Nasir, M.; Gupta, A.; Beg, M.; Chua, G.K.; Kumar, A. Physical and mechanical properties of medium density fibreboard using soy-lignin adhesives. *J. Trop For. Sci.* **2014**, *1*, 41–49.

40. Papadopoulou, E.; Kountouras, S.; Chrissafis, K.; Kirpluks, M.; Cabulis, U.; Šviglerova, P.; Benjelloun-Mlayah, B. Evaluation of the particle size of organosolv lignin in the synthesis of resol resins for plywood and their performance on fire spreading. *TAPPI J.* **2017**, *16*, 409–416. [CrossRef]

41. Sepahvand, S.; Doosthosseini, K.; Pirayesh, H.; Maryan, B.K. Supplementation of natural tannins as an alternative to formaldehyde in urea and melamine formaldehyde resins used in MDF production. *Drv. Ind.* **2018**, *69*, 215–221. [CrossRef]

42. Hosseinpourpia, R.; Adamopoulos, S.; Mai, C.; Taghiyari, H.R. Properties of medium-density fibreboards bonded with dextrin-based wood adhesives. *Wood Res.* **2019**, *64*, 185–194.

43. Tisserat, B.; Eller, F.; Mankowski, E. Properties of composite wood panels fabricated from Eastern red cedar employing various bio-based green adhesives. *BioResources* **2019**, *14*, 6666–6685.

44. Lora, J.H.; Glasser, W.G. Recent industrial applications of lignin: A sustainable alternative to nonrenewable materials. *J. Polym. Environ.* **2002**, *10*, 39–48. [CrossRef]

45. Glasser, W.G.; Kelly, S.S. Lignin. In *Encyclopedia of Polymer Science and Engineering*, 2nd ed.; Mark, H.F., Bikales, N., Overberger, C.G., Menges, G., Kroschwitz, J.I., Eds.; Wiley: New York, NY, USA, 1987; pp. 795–852.

46. Argyropoulos, D.S.; Menachem, S.B.; Eriksson, K.E.L.; Babel, W.; Blanch, H.W.; Cooney, C.L.; Enfors, S.O.; Eriksson, K.E.L. (Eds.) *Biotechnology in the Pulp and Paper Industry*; Springer: Berlin/Heidelberg, Germany, 1997; pp. 127–158.

47. Fu, D.; Mazza, G.; Tamaki, Y. Lignin extraction from straw by ionic liquids and enzymatic hydrolysis of the cellulosic residues. *J. Agric. Food Chem.* **2010**, *58*, 2915–2922. [CrossRef] [PubMed]

48. Pizzi, A. Bioadhesives for wood and fibres. *Rev. Adhes. Adhes.* **2013**, *1*, 88–113. [CrossRef]

49. Laurichesse, S.; Avérous, L. Chemical modification of lignins: Towards biobased polymers. *Prog. Polym. Sci.* **2014**, *39*, 1266–1290. [CrossRef]

50. Chakar, F.S.; Ragauskas, A.J. Review of current and future softwood kraft lignin process chemistry. *Ind. Crops Prod.* **2004**, *20*, 131–141. [CrossRef]

51. Gargulak, J.D.; Lebo, S.E. Commercial use of lignin-based materials. In *Lignin: Historical, Biological, and Material Perspectives*; Glasser, W.G., Northey, R.A., Schultz, T.P., Eds.; American Chemical Society: Washington, DC, USA, 2000; pp. 304–320.

52. Waldron, K. *Advances in Biorefineries: Biomass and Waste Supply Chain Exploitation*; Woodhead Publishing: London, UK, 2014.

53. Vanholme, R.; Demedts, B.; Morreel, K.; Ralph, J.; Boerjan, W. Lignin biosynthesis and structure. *Plant. Physiol.* **2016**, *153*, 895–905. [CrossRef]

54. Geng, X.; Li, K. Investigation of wood adhesives from kraft lignin and polyethylenimine. *J. Adhes. Sci. Technol.* **2006**, *20*, 847–858. [CrossRef]

55. Mancera, C.; El Mansouri, N.E.; Vilaseca, F.; Ferrando, F.; Salvado, J. The effect of lignin as a natural adhesive on the physico-mechanical properties of *Vitis vinifera* fibreboards. *BioResources* **2011**, *6*, 2851–2860.

56. Ghaffar, S.H.; Fan, M. Lignin in straw and its applications as an adhesive. *Int. J. Adhes. Adhes.* **2014**, *48*, 92–101. [CrossRef]

57. Yotov, N.; Valchev, I.; Petrin, S.; Savov, V. Lignosulphonate and waste technical hydrolysis lignin as adhesives for eco-friendly fibreboard. *Bulg. Chem. Commun.* **2017**, *49*, 92–97.

58. Klapiszewski, Ł.; Jamrozik, A.; Strzemiecka, B.; Voelkel, A.; Jesionowski, T. Activation of magnesium lignosulfonate and kraft lignin: Influence on the properties of phenolic resin-based composites for potential applications in abrasive materials. *Int. J. Mol. Sci.* **2017**, *18*, 1224. [CrossRef] [PubMed]

59. Klapiszewski, Ł.; Oliwa, R.; Oleksy, M.; Jesionowski, T. Calcium lignosulfonate as eco-friendly additive of crosslinking fibrous composites with phenol-formaldehyde resin matrix. *Polymers* **2018**, *63*, 102–108. [CrossRef]

60. Hazwan Hussin, M.; Aziz, A.A.; Iqbal, A.; Ibrahim, M.; Latif, N. Development and characterization novel bio-adhesive for wood using kenaf core (*Hibiscus cannabinus*) lignin and glyoxal. *Int. J. Biol. Macromol.* **2019**, *122*, 713–722. [CrossRef] [PubMed]

61. Zhang, J.; Wang, W.; Zhou, X.; Liang, J.; Du, G.; Wu, Z. Lignin-based adhesive cross-linked by furfuryl alcohol-glyoxal and epoxy resins. *Nord. Pulp. Paper Res. J.* **2019**, *34*, 228–238. [CrossRef]

62. Vázquez, G.; González, J.; Freire, S.; Antorrena, G. Effect of chemical modification of lignin on the gluebond performance of lignin-phenolic resins. *Biores. Technol.* **1997**, *60*, 191–198. [CrossRef]

63. Fan, J.; Zhan, H. Optimization of synthesis of spherical lignosulphonate resin and its structure characterization. *Chin. J. Chem. Eng.* **2008**, *16*, 407–410. [CrossRef]

64. Vishtal, A.; Kraslawski, A. Challenges in industrial applications of technical lignins. *BioResources* **2011**, *6*, 3547–3568.

65. Lora, J. Industrial commercial lignins: Sources, properties and applications. In *Monomers, Polymers and Composites from Renewable Resources*; Belgacem, M.N., Gandini, A., Eds.; Elsevier: Amsterdam, The Netherlands, 2008; pp. 225–241.

66. El Mansouri, N.E.; Salvadó, J. Structural characterization of technical lignins for the production of adhesives: Application to lignosulphonate, kraft, soda- anthraquinone, organosolv and ethanol process lignins. *Ind. Crops Prod.* **2006**, *24*, 8–16. [CrossRef]

67. Miller, J.; Faleiros, M.; Pilla, L.; Bodart, A.C. *Lignin: Technology, Applications and Markets, Special Market Analysis Study*; RISI, Inc., Market-Intell LCC: Charlottesville, VA, USA, 2016.

68. Ibrahim, M.N.M.; Zakaria, N.; Sipaut, C.S.; Sulaiman, O.; Hashim, R. Chemical and thermal properties of lignins from oil palm biomass as a substitute for phenol in a phenol formaldehyde resin production. *Carbohydr. Polym.* **2011**, *86*, 112–119. [CrossRef]

69. Podschun, J.; Stücker, A.; Buchholz, R.I.; Heitmann, M.; Schreiber, A.; Saake, B.; Lehnen, R. Phenolated lignins as reactive precursors in wood veneer and particleboard adhesion. *Ind. Eng. Chem. Res.* **2016**, *55*, 5231–5237. [CrossRef]

70. Hemmilä, V.; Adamopoulos, S.; Hosseinpourpia, R.; Sheikh, A.A. Ammonium lignosulfonate adhesives for particleboards with pMDI and furfuryl alcohol as cross-linkers. *Polymers* **2019**, *11*, 1633. [CrossRef] [PubMed]

71. Antov, P.; Savov, V.; Mantanis, G.I.; Neykov, N. Medium-density fibreboards bonded with phenol-formaldehyde resin and calcium lignosulfonate as an eco-friendly additive. *Wood Mater. Sci. Eng.* **2020**. [CrossRef]

72. Danielson, B.; Simonson, R. Kraft lignin in phenol formaldehyde resin. Part 1. Partial replacement of phenol by kraft lignin in phenol formaldehyde adhesives for plywood. *J. Adhes. Sci. Technol.* **1998**, *12*, 923–939. [CrossRef]

73. Kumar, R.N.; Pizzi, A. Thermosetting Adhesives Based on Bio-Resources for Lignocellulosic Composites. In *Adhesives for Wood and Lignocellulosic Materials*; Wiley-Scrivener Publishing: Hoboken, NJ, USA, 2019; pp. 275–277.

74. Antov, P.; Valchev, I.; Savov, V. Experimental and statistical modeling of the exploitation properties of eco-friendly MDF through variation of lignosulfonate concentration and hot pressing temperature. In Proceedings of the 2nd International Congress of Biorefinery of Lignocellulosic Materials (IWBLCM2019), Cordoba, Spain, 4–7 June 2019; pp. 104–109, ISBN 978-84-940063-7-1.

75. Savov, V.; Valchev, I.; Antov, P. Processing factors affecting the exploitation properties of environmentally friendly medium density fibreboards based on lignosulfonate adhesives. In Proceedings of the 2nd International Congress of Biorefinery of Lignocellulosic Materials (IWBLCM2019), Cordoba, Spain, 4–7 June 2019; pp. 165–169, ISBN 978-84-940063-7-1.

76. European Committee for Standardization, EN 310. *Wood-Based Panels-Determination of Modulus of Elasticity in Bending and of Bending Strength*; European Committee for Standardization: Brussels, Belgium, 1999.

77. European Committee for Standardization, EN 317. *Particleboards and Fibreboards-Determination of Swelling in Thickness after Immersion in Water*; European Committee for Standardization: Brussels, Belgium, 1998.

78. European Committee for Standardization, EN 322. *Wood-Based Panels-Determination of Moisture Content*; European Committee for Standardization: Brussels, Belgium, 1998.

79. European Committee for Standardization, EN 323. *Wood-Based Panels-Determination of Density*; European Committee for Standardization: Brussels, Belgium, 2001.

80. European Committee for Standardization, EN ISO 12460-5. *Wood-Based Panels-Determination of Formaldehyde Release–Part 5. Extraction Method (Called the Perforator Method)*; European Committee for Standardization: Brussels, Belgium, 2015.

81. Salem, M.Z.M.; Böhm, M. Understanding of formaldehyde emissions from solid wood: An overview. *BioResources* **2013**, *8*, 4775–4790. [CrossRef]

82. Schäfer, M.; Roffael, E. On the formaldehyde release of wood. *Holz. Roh. Werkst.* **2000**, *58*, 259–264. [CrossRef]

83. European Committee for Standardization, EN 312. *Particleboards–Specifications*; European Committee for Standardization: Brussels, Belgium, 2010.

84. European Committee for Standardization, EN 622-5. *Fibreboards-Specifications-Part 5: Requirements for Dry Process Boards*; European Committee for Standardization: Brussels, Belgium, 2010.

85. Shiag, K.; Negi, A. Physical and mechanical properties of MDF using needle punching technique. *Int. J. Chem. Sci.* **2017**, *5*, 2028–2030.

Article

The Impact of a CO_2 Laser on the Adhesion and Mold Resistance of a Synthetic Polymer Layer on a Wood Surface

Ladislav Reinprecht and Zuzana Vidholdová

Department of Wood Technology, Faculty of Wood Sciences and Technology, Technical University in Zvolen, T. G. Masaryka 24, SK-960 01 Zvolen, Slovakia; zuzana.vidholdova@tuzvo.sk
* Correspondence: reinprecht@tuzvo.sk; Tel.: +421-45-520-6383

Abstract: In the wood industry, laser technologies are commonly applied for the sawing, engraving, or perforation of solid wood and wood composites, but less knowledge exists about their effect on the joining and painting of wood materials with synthetic polymer adhesives and coatings. In this work, a CO_2 laser with irradiation doses from 2.1 to 18.8 J·cm^{-2} was used for the modification of European beech (*Fagus sylvatica* L.) and Norway spruce (*Picea abies* /L./ Karst) wood surfaces—either in the native state or after covering them with a layer of polyvinyl acetate (PVAc) or polyurethane (PUR) polymer. The adhesion strength of the phase interface "synthetic polymer—wood", evaluated by the standard EN ISO 4624, decreased significantly and proportionately in all the laser modification modes, with higher irradiation doses leading to a more apparent degradation and carbonization of the wood adherent or the synthetic polymer layer. The mold resistance of the polymers, evaluated by the standard EN 15457, increased significantly for the less mold-resistant PVAc polymer after its irradiation on the wood adherent. However, the more mold-resistant PUR polymer was able to better resist the microscopic fungi *Aspergillus niger* Tiegh. and *Penicillium purpurogenum* Stoll. when irradiation doses of higher intensity acted firstly on the wood adherent.

Keywords: wood; polyvinyl acetate; polyurethane; laser; damage; microscopy; adhesion; mold

Citation: Reinprecht, L.; Vidholdová, Z. The Impact of a CO_2 Laser on the Adhesion and Mold Resistance of a Synthetic Polymer Layer on a Wood Surface. *Forests* **2021**, *12*, 242. https://doi.org/10.3390/f12020242

Academic Editor: Vikram Yadama

Received: 25 January 2021
Accepted: 17 February 2021
Published: 20 February 2021

Publisher's Note: MDPI stays neutral with regard to jurisdictional claims in published maps and institutional affiliations.

1. Introduction

The intentional processing or modification of wood surfaces can be performed by various mechanical, physical, chemical, and biological methods. Their aim is to improve or optimize (1) the final characteristics of wood surfaces—e.g., their shape, roughness, hardness, color, gloss, and resistance (to water, sun, fire, fungi and other biological agents) [1–4]; (2) the inter-operational characteristics of wood surfaces—e.g., the wettability, free surface energy, and density—which are important before wood gluing/painting with natural and synthetic polymer adhesives/coatings [5–7]. A polymer layer of the adhesive/coating can also be processed or modified with the aim to achieve its general or specific optimum properties either for the final wood product (e.g., veneers on surface-treated wood furniture) [8] or for the further wood processing (e.g., the gluing of solid wood or wood particles in wood composites) [9].

Several physical methods can be used for the surface modification of wood in its native state as well as after its pretreatment with biocides, UV-absorbers, coatings, and other chemicals [2,10]. Today, the most perspective physical methods for wood surface modification are plasma methods [11–14] and laser methods [15–19]. Both these physical methods change the chemical structure of wood and other organic materials [2,17,20]. In this context, they should be able to change several characteristics of the wood surface and also of the adhesive or coating polymer layer applied on the wood surface-including the final adhesion strength and mold resistance.

Plasma changes the chemical composition of wood surfaces and also their wettability and some other properties [1,11,12,21]. Plasma in the presence of air in the lignin-polysaccharide matrix of wood specifically generates new hydroxyl, carbonyl, carboxyl, peroxide or ether functional groups, and radicals, which are able to enter into further condensation reactions. Thereby, plasma on the wood surfaces forms a thin layer which is more wettable for polar liquids, with a consequent positive effect on the adhesion strength of polar and semi-polar coatings or adhesives to wood [12,13,22].

Laser technologies in the wood industry are commonly applied for the cutting, engraving, drilling/incising, surface treatment, or cleaning of solid wood and wood composites [23–30]. Laser beams change the chemical structure of irradiated wood surfaces [9,31]; their anatomy and morphology [23,25,32–34]; as well as their properties—e.g., their color, roughness, wettability, free surface energy, and resistance (to bacteria, molds, and other biological agents) [4,18,28,35,36]. The results of the X-ray photo-electron spectroscopy indicated an increased number of non-polar bonds such as C–C and C–H, with C–O bonds maintained without change [33]. The microscopic observations point out that treating wood surfaces with a laser beam may make the surface smoother because the cells melt down to a depth of several micrometers without their direct carbonization [33]. Dolan et al. [37] observed that cellulose fibrils coalesced and formed a bubble-like topology when exposed to the energy of a CO_2 laser. A laser is able to create similar changes also in other natural and synthetic polymer materials in connection with degradation, depolymerization, and carbonization processes in various types of organic macromolecules [38]. Lasers, unlike plasma, hydrophobize wood components and worsen the wettability of wood with water and other polar liquids-including polar synthetic polymer adhesives and coatings [6,12].

Adhesion processes play a significant role in contemporary material bonding and painting technologies. The final quality of bonded or painted wooden materials is conditioned by the interaction between the wood substrate and the liquid and following solid-cured adhesives or coatings. The formation of the "adhesive/coating–wood" phase interface is significantly determined mainly by two factors: (1) the surface properties of the wood adherent (e.g., density, porosity, roughness, wettability, free surface energy, pH value) and (2) the physical/chemical and application properties of the used adhesive or coating (e.g., polymer type, molecular weight, viscosity, surface tension, polarity, pH value, weight solid, density). However, some other factors can also affect the adhesion processes of wood bonding and painting—for example, the initial chemical, physical, or biological pretreatments of wood surfaces or the creation of internal stresses in the deposited polymer films or in the created "adhesive/coating–wood" interfaces. Surface pretreatment is one of the first and most important technological stages in the adhesive bonding process. Ülker [39] underlines that a particular pretreatment of wood for structural bonding will ideally produce a surface which is (1) free from contamination, (2) wettable by an adhesive, (3) optimally macro- and micro-rough, (4) mechanically stable, and (5) hydrolytically stable.

The biological resistance of bonded and painted wood against bacteria, molds, staining fungi, and decaying fungi is important mainly for products exposed to a moist environment [40,41]. Several natural adhesives and coatings—e.g., those based on sugar and protein macromolecules—are easily attacked and destroyed by biological agents. Such bio-degradations can then be quickly connected with the delamination of bonded wood products or the peeling of coatings from painted wood surfaces. Synthetic polymers have usually a higher resistance to biological attacks, and the most resistant to bacteria and molds are the polyurethane (PUR) polymers [42,43].

Assembly adhesives and coatings recommended for wooden products, such as solid wood panels, glulam, furniture parts, and sport or musical instruments, are manufactured from several synthetic polymers, including polyvinyl acetate (PVAc) and PUR polymers [44].

The PVAc polymer is the most widely used as a water emulsion. Its emulsion consistency is created by the polymerization of polar vinyl monomers, predominantly vinyl acetate, in water. From PVAc polymers, thermoplastic adhesives that are important for furniture manufacturing and carpentry are produced. The bonding principle of PVAc adhesives is based on the removal of water by penetration into the wood substrate or by the evaporation of water to the surrounding air. The forming of the bond also requires the application of proper pressure. PVAc polymer systems show a long storage life and are easy to clean. However, the bond formed from them has a lower resistance to heat and is characterized by creep behavior. The biodeterioration of PVAc polymers occurs due to the enzymatic action of esterase created by filamentous fungi, whereas algae, yeasts, lichens, and bacteria can also degrade polyvinyl acetate macromolecules [45].

PUR polymers are synthetized by the reaction of various types of isocyanates with polyols, forming repeating polar urethane moiety units which have a good bonding ability to various surfaces. The biodeterioration of PUR polymers can occur through the enzymatic action of various hydrolases, such as ureases, proteases, and esterases, which are created by fungi (e.g., *Chaetomium globosum* Kunze ex Fr.) and bacteria (e.g., *Bacillus subtilis* Cohn) [46].

The aim of this work was to analyze the effect of a CO_2 laser on the adhesion strength and the resistance to molds of the PVAc and PUR polymers applied to the surfaces of beech and spruce woods.

2. Materials and Methods

2.1. Woods

The samples with dimensions of $250 \times 130 \times 10$ mm (longitudinal $\times$ radial $\times$ tangential) were prepared from the boards of two wood species—European beech (*Fagus sylvatica* L.) and Norway spruce (*Picea abies* /L./Karst)—after their 2-year air seasoning under the roof. In total, 76 wood samples were collected for the experiment, including samples without knots, false heartwood, biological damage, growth defects, or inhomogeneity. Before performing the individual technological operations—(1) the application of the synthetic polymer layer directly on the top surface of the native wood or (2) the laser modification of the top surface of the native wood—the top surfaces of the native wood samples were sanded along the fibers with sandpapers of 80 grit and then 120 grit on the belt DREMEL®3000 grinder machine (Breda, The Netherlands). Following this, the wood samples were conditioned for 7 days at a temperature of 20 °C and a relative air humidity of 65%, achieving a moisture content from 11% to 13%.

2.2. Synthetic Polymers

The PVAc polymer was used in the form of a one-component water-dispersed adhesive TechnoBond D3 P (Agglu, Turčianske Teplice, Slovakia); see Table 1.

Table 1. Synthetic polymers in commercial adhesives and their processing conditions.

Synthetic Polymer	Polyvinyl Acetate PVAc	Polyurethane PUR
Adhesive	TechnoBond D3 P	Neopur 1791 and Adiflex 935
Density (kg·m^{-3})	1080	1550
Weight solids (%)	51 ± 2	100
Colour after curing	transparent	light-white
pH value	3 ± 0.4	-
Spread rate (g·m^{-2})	120–200	200–400
Open time (min)	15	90–130
Pressing time at 20 °C (h)	0.3	5

The PUR polymer was applied as two-component reactive adhesive Neopur 1791 mixed before the application of the hardener Adiflex 935 in a weight ratio of 100:20 (Neoflex Adhesives, Elche, Spain)—Table 1.

Both polymers in the form of adhesives were applied to the top surfaces of samples conditioned at a moisture content of $12 \pm 1\%$ by a hand toothed trowel in an amount (spread rate) of 200 g·m^{-2} before or after the irradiation of the samples with a CO_2 laser. Following this, the samples were conditioned for 7 days without pressing at a temperature of 20 °C and a relative air humidity of 65%, achieving a moisture content from 11% to 13%.

2.3. Laser Irradiation of Surfaces

The commercial CO_2 laser LCS 100 device (Formetal Piesok, Piesok, Slovakia), working with a wavelength of 10.6 µm, was used for the irradiation of the top surfaces of the native wood samples as well as of the wood samples first covered with a layer of synthetic polymer. The effective power of the laser beam (45 W) was determined in continuous-wave mode using the equipment Coherent Radiation Model 201 (Coherent Inc., Santa Clara, CA, USA). The laser beam diameter ($\varnothing$), used for the irradiation of the sample in one pass, was constantly 12 mm. The laser head moved perpendicularly to the top surface of sample at a determined scanning speed v from 18 to 2 cm·s^{-1} (Table 2). With the aim to obtain an approximately uniform intensity of the irradiation dose H across the whole surface of the sample, with a width of 130 mm in the radial direction, the laser head moved in twenty-two passes parallel with the longitudinal direction of the sample, overlapping homogenously with its surface (Figure 1).

Table 2. The laser irradiation doses (H) acting on the top surface of samples dependent on the scanning speed (v).

Sample	Ref.	A	B	C	D	E	F	G	H	I
Scanning speed v (cm·s^{-1})	-	18	16	14	12	10	8	6	4	2
Irradiation dose H (J·cm^{-2})	0	2.1	2.3	2.7	3.1	3.8	4.7	6.3	9.4	18.8

Notes: Ref. represents reference samples (in total, 4 types of reference samples were used—i.e., from 2 wood species "beech or spruce" covered with 2 polymer types "PVAc or PUR"). Symbols A to I represent testing samples modified with 9 different irradiation doses (in total, 72 types of laser-irradiated samples—i.e., 9 irradiation doses "from A to I" × 2 wood species "beech or spruce" × 2 polymer types "PVAc or PUR" × 2 modes of polymer application "before or after irradiation of sample with a CO_2 laser").

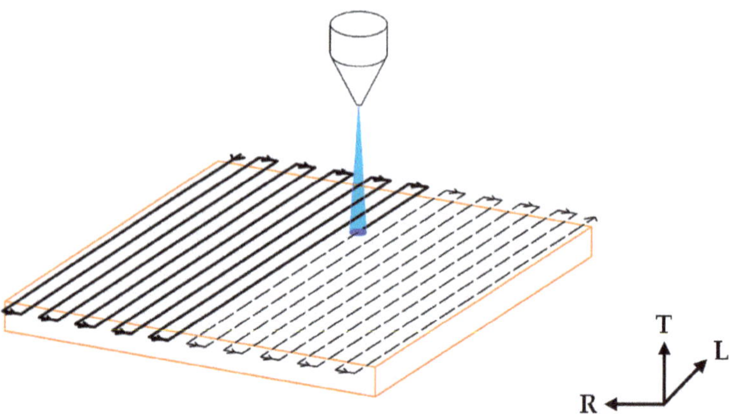

Figure 1. Scheme of the sample 250 × 130 × 10 mm (longitudinal × radial × tangential) irradiated with the beam of a CO_2 laser.

The irradiation doses H (J·cm^{-2}) were calculated for the individual scanning speeds of the laser head v (cm·s^{-1}) and the other laser parameters in the modification process according to Equation (1):

$$H = (Pe \times \tau): S = (Pe \times L): (S \times v) \ (\text{J·cm}^{-2}),\tag{1}$$

where Pe (45 W) is the effective laser beam power acting on the top surface of the sample, τ (s) is the irradiation time of one pass (i.e., the ratio between the dimension of the wood sample in the direction of its irradiation L (25 cm) and the scanning speed of the laser head v (from 18 cm·s^{-1} to 2 cm·s^{-1})), and S ($\varnothing \times L = 1.2$ cm × 25 cm = 30 cm^2) is the area irradiated during one pass (i.e., an area equal to the homogenously irradiated area by the Gaussian distribution of the real laser intensity distribution for this area at three passes) (Figure 1).

2.4. Light Microscopy Analysis

The effects of a CO_2 laser on the structural changes in the top surfaces of wood samples covered with a layer of the synthetic polymers were assessed with the light microscope Olympus BX43F (Olympus Corporation, Tokyo, Japan) at a magnification of $10\times$.

2.5. Adhesion Strength of the "Synthetic Polymer—Wood" Interface

The adhesion strength (σ) of the synthetic polymer layer to the wood adherent was determined by the pull-off test for adhesion in accordance with the standard EN ISO 4624 [47], using the PosiTest AT-M Adhesion Tester instrument (DeFelsko, Ogdensburg, NY, USA). A bond connection of the steel-roller dolly ($\varnothing = 20$ mm) with the top surface of the sample (polymer layer present on wood adherent) was made by two-component epoxy resin (Repair Universal Epoxy Resin, Pattex, Henckel, Germany).

The pull-off test tensile strength method measures the tensile force perpendicular to the phase interface "adhesive/coating–wood" system required to tear off the steel-roller from the adhesive or coating present on the surface of the wood sample, at which the failures could occur either in the weakest interface "adhesive/coating–wood" or in the weakest component "adhesive/coating" or "wood".

For each sample, the adhesion strength was determined in 4 places (Figure 2). In total, 304 adhesion tests were performed for 76 samples—i.e., for 4 reference samples and 72 samples modified with a CO_2 laser (i.e., 2 wood species: beech and spruce; 2 polymer types: PVAc and PUR; 2 modes of CO_2 laser modification: sample irradiated before or after covering with a layer of polymer; 9 laser irradiation doses. See Table 1).

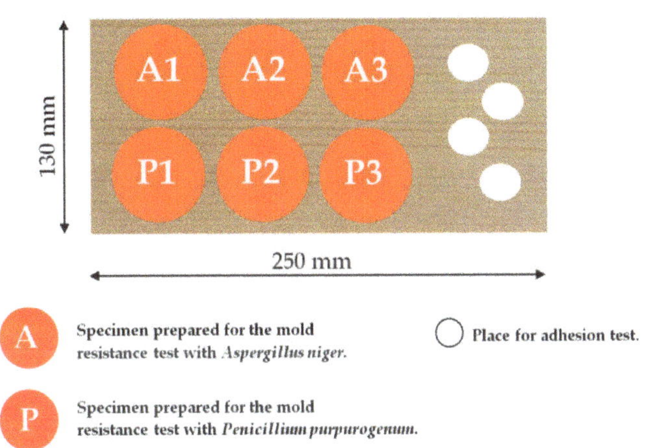

Figure 2. Scheme of sample with 4 places for the adhesion test and with 6 specimens prepared from it for the mold resistance test.

2.6. Mold Resistance of the Synthetic Polymer Layer on the Wood Adherent

For the mold growth activity (MGA) test, the top surfaces of circular specimens with a diameter of 54 mm (prepared from the relevant sample; see Figure 2) were exposed in accordance with the standard EN 15457 [48]. Sterilized specimens were placed into Petri dishes with a diameter of 120 mm on a 3 to 4 mm-thick layer of 4.9 wt.% Czapek–Dox agar medium (HiMedia Ltd., Mumbai, India) and inoculated with a spore suspension of the microscopic fungi *Aspergillus niger* Tiegh. (strain BAM 122; Bundesanstalt für Materialforschung und -prüfung, Berlin) and *Penicillium purpurogenum* Stoll. (strain BAM

24; Bundesanstalt für Materialforschung und -prüfung, Berlin), respectively. Both spore suspensions were prepared in sterile water in concentrations of 10^6 to 10^7 spores/mL. The incubation of the inoculated specimens lasted 21 days at a temperature of 24 °C $\pm$ 2 °C and a relative air humidity of 90% to 95%. The mold resistance of the specimens was determined by the *MGA* values (from 0 to 4) using these criteria: 0, no mold growth on the top surface; 1, $\leq$10%; 2, >10% but $\leq$30%; 3, >30% but $\leq$50%; 4, >50%.

2.7. Statistical Analysis

The measured data of the adhesion strength (σ) and the mold growth activity (*MGA*) were for individual groups of the reference and laser-modified samples evaluated on the basis of mean values and standard deviations. Using ANOVA, the effect of increased laser irradiation doses (*H*) from 2.1 to 18.8 J·cm^{-2} on the adhesion strength and the mold resistance was analyzed by the linear correlations "$\sigma = a + b \cdot H$" and "$MGA = a + b \cdot H$", together with the coefficients of determination, r^2, and levels of significance, *p*.

3. Results and Discussion

3.1. Microscopy of Surfaces

The synthetic polymer layers present on the wood surfaces showed visible structural changes after irradiation with a CO_2 laser (Figures 3 and 4).

Due to the relatively smaller irradiation dose of 4.7 J·cm^{-2}, in the layers of PVAc and PUR polymers present on the wood surfaces air bubbles with diameters of 0.1 to 0.3 mm were created (Figure 3c,d and Figure 4c,d). The PVAc polymer obtained a yellow shade as well (Figure 3c,d). The bubbles in the synthetic polymer layers probably arose as a result of thermal decomposition reactions in their macromolecules, together with the creation of lower-molecular gasses which inflated a layer of thermoplastic PVAc polymer or a layer of the completely non-crosslinked PUR polymer. Similar bubbles in layers of synthetic polymers are intentionally created in intumescent types of fire retardants [49,50]. However, bubbles created due to CO_2 lasers could also be water molecules released from moist wood adherents (i.e., the water bonded in wood by hydrogen and other physical-chemical forces), as well as water and various small organic molecules created during the thermal destruction of the lignin-polysaccharide matrix of wood. A yellow shade in the laser-irradiated PVAc polymer could be explained by its lower thermal stability [51,52] in comparison to the PUR polymer [53,54].

By applying the highest irradiation dose of 18.8 J·cm^{-2}, the layers of PVAc and PUR polymers were evidently carbonized. They obtained brown shades until they were nearly a black color, at which the bubbles in the carbonized polymer layers had already burst in most cases (Figure 3e,f and Figure 4e,f).

The primary laser irradiation of the native wood surfaces also caused decomposition and carbonization processes—in this case, in the lignin-polysaccharide components of wood. The irradiation of wood surfaces with higher doses *H* was connected with their darkening—gradually from yellowing to browning and blackening—in accordance with other research works—e.g., [5,6,17–19,35].

3.2. Adhesion Strength

For the reference samples—i.e., samples unmodified with a CO_2 laser, a better adhesion strength between the layer of synthetic polymer and the wood adherent was determined for the PUR polymer (beech: σ = 5.33 MPa; spruce: σ = 3.05 MPa) than for the PVAc polymer (beech: σ = 3.69 MPa; spruce: σ = 2.35 MPa) (Tables 3 and 4).

The adhesion strength (σ) of the "synthetic polymer–wood" phase interface continuously decreased with the increase in the irradiation dose (*H*) of a CO_2 laser acting on the top surface of wood samples before or after covering them with a layer of synthetic polymer (Tables 3 and 4, Figures 5–7).

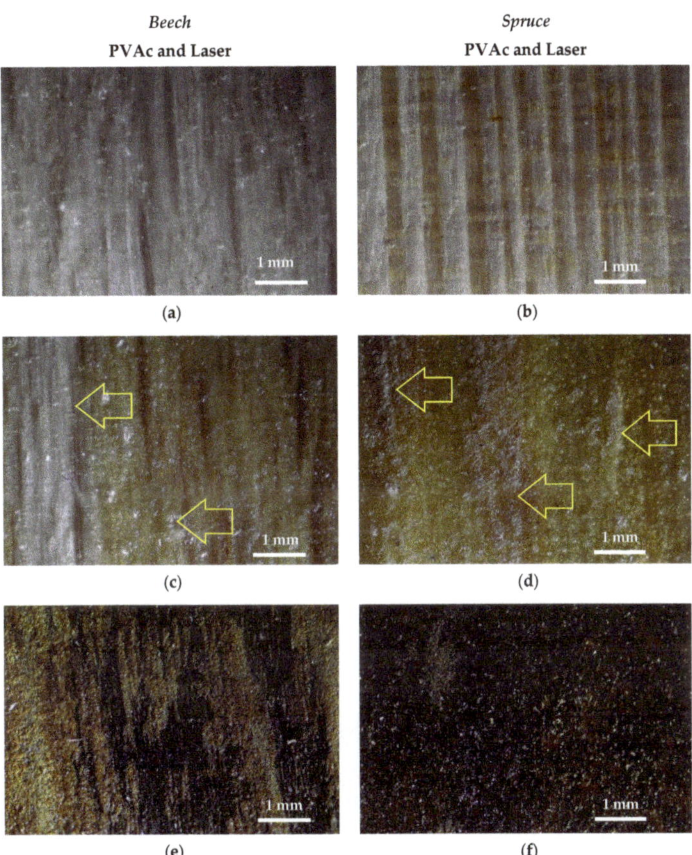

Beech
PVAc and Laser   *Spruce*
PVAc and Laser

(a)   (b)

(c)   (d)

(e)   (f)

Figure 3. The microscopic image of the top surfaces of wood samples covered with a layer of PVAc polymer—reference (**a,b:** H = 0.0 J·cm^{-2}) and irradiated with a CO_2 laser with the mode "PVAc and Laser" when applying a medium irradiation dose (**c,d:** H = 4.7 J·cm^{-2}) with the irregular occurrence of air bubble zones in the heat-affected polymer layers, or a maximum irradiation dose (**e,f:** H = 18.8 J·cm^{-2}) with the intensive creation of air bubbles in the strongly carbonized polymer layers.

Table 3. The adhesion strength (σ) of the "synthetic polymer–wood" interface for the European beech wood samples irradiated with various doses of CO_2 laser (H) before or after application of the PVAc or PUR polymer layer.

Irradiation Dose H (J·cm^{-2})	Ref. 0	A 2.1	B 2.3	C 2.7	D 3.1	E 3.8	F 4.7	G 6.3	H 9.4	I 18.8
Beech	Adhesion strength—σ (MPa)									
Laser and PVAc	3.69 (0.73)	3.33 (0.46)	3.04 (0.39)	3.24 (0.29)	2.34 (0.36)	2.33 (0.41)	2.91 (0.71)	2.25 (0.45)	2.46 (0.62)	2.09 (0.27)
PVAc and Laser	3.69 (0.73)	3.01 (0.74)	3.18 (0.49)	2.84 (0.38)	2.45 (0.15)	2.32 (0.50)	2.77 (0.47)	2.49 (0.44)	2.35 (0.63)	1.95 (0.36)
Laser and PUR	5.33 (0.61)	3.52 (0.78)	2.77 (0.30)	2.79 (0.79)	3.08 (0.51)	2.19 (0.11)	2.25 (0.42)	2.73 (0.88)	2.00 (0.63)	1.59 (0.57)
PUR and Laser	5.33 (0.61)	4.69 (0.95)	4.57 (0.99)	3.89 (1.08)	3.84 (0.69)	2.90 (0.97)	2.87 (0.77)	3.17 (1.17)	2.87 (0.90)	2.03 (0.59)

Note: Mean value is from 4 measurements. Standard deviation is in parentheses.

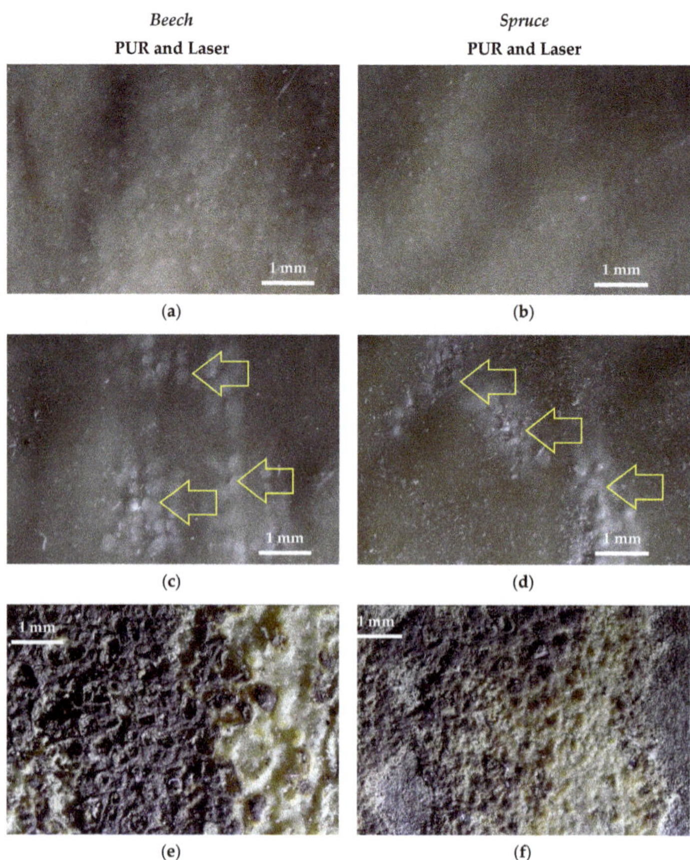

Figure 4. The microscopic image of the top surfaces of wood samples covered with a layer of PUR polymer—reference (**a,b**: H = 0.0 J·cm^{-2}) and irradiated with a CO_2 laser with the mode "PUR and Laser" when applying a medium irradiation dose (**c,d**: H = 4.7 J·cm^{-2}) with the irregular occurrence of air bubble zones in the heat-affected polymer layers, or a maximum irradiation dose (**e,f**: H = 18.8 J·cm^{-2}) with the intensive creation of air bubbles and their bursting in the strongly carbonized polymer layers.

Table 4. The adhesion strength (σ) of the "synthetic polymer–wood" interface for the Norway spruce wood samples irradiated with various doses of CO_2 laser (H) before or after the application of the PVAc or PUR polymer layer.

Irradiation Dose H ($J·cm^{-2}$)	Ref. 0	A 2.1	B 2.3	C 2.7	D 3.1	E 3.8	F 4.7	G 6.3	H 9.4	I 18.8
Spruce	Adhesion strength—σ (MPa)									
Laser and PVAc	2.35 (0.53)	2.27 (0.41)	2.20 (0.39)	1.97 (0.08)	2.08 (0.69)	1.93 (0.81)	1.78 (0.22)	1.78 (0.28)	1.18 (0.21)	1.22 (0.22)
PVAc and Laser	2.35 (0.53)	2.09 (0.26)	2.15 (0.42)	1.99 (0.12)	1.93 (0.40)	2.02 (0.29)	1.75 (0.34)	1.41 (0.25)	1.38 (0.12)	1.45 (0.26)
Laser and PUR	3.05 (0.38)	2.41 (0.41)	2.29 (0.58)	1.93 (0.37)	1.72 (0.51)	1.91 (0.24)	1.84 (0.25)	1.74 (0.52)	1.34 (0.40)	1.50 (0.25)
PUR and Laser	3.05 (0.38)	2.20 (0.50)	1.95 (0.57)	1.97 (0.50)	1.83 (0.76)	1.59 (0.30)	1.79 (0.28)	1.51 (0.23)	1.57 (0.33)	1.53 (0.21)

Note: Mean value is from 4 measurements. Standard deviation is in parentheses.

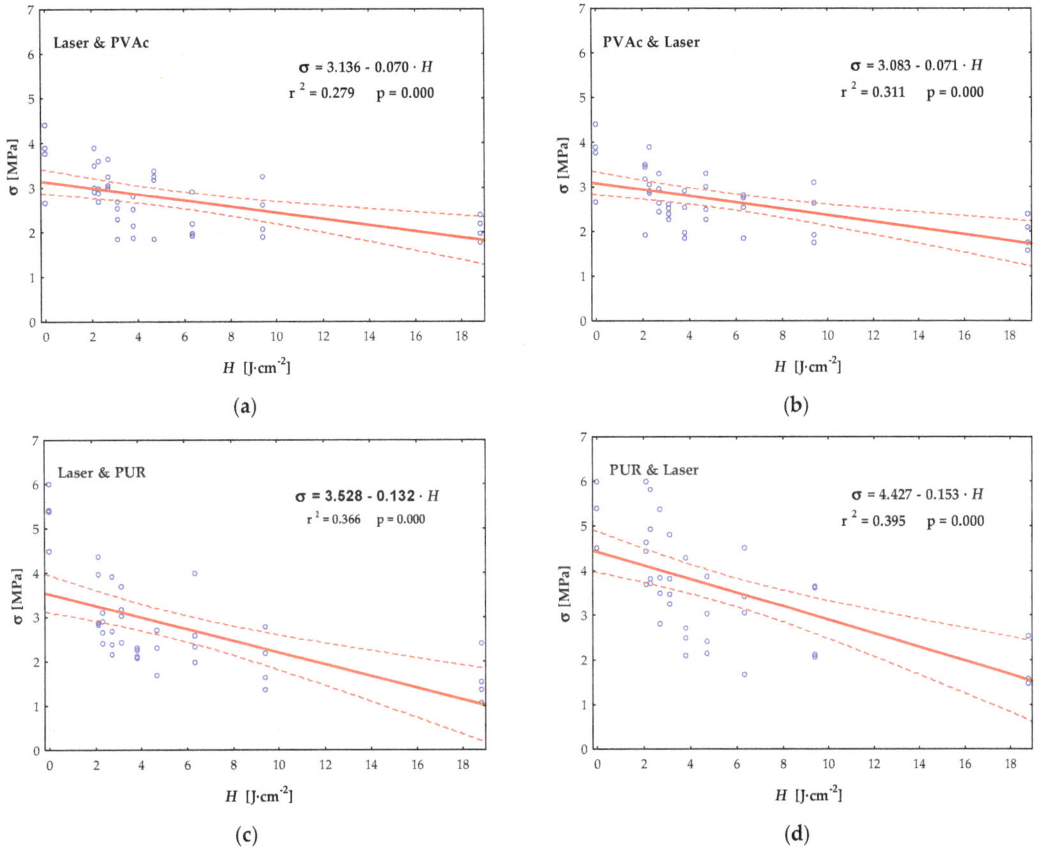

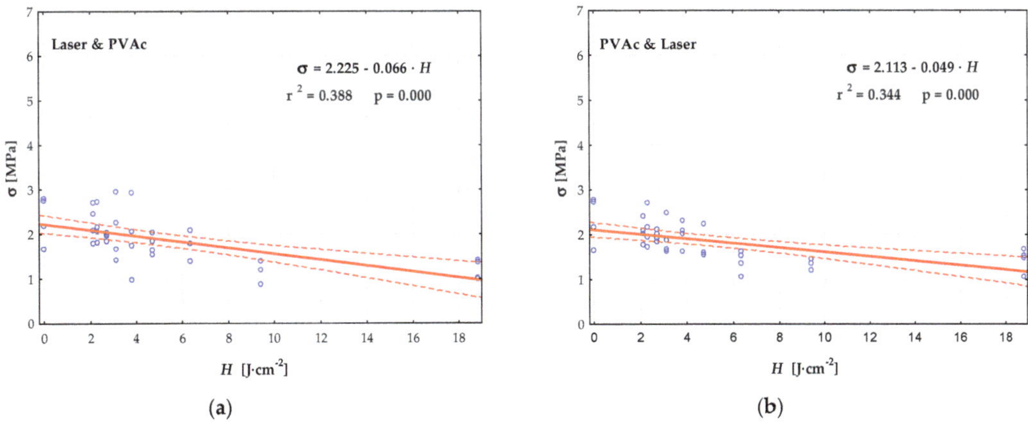

Figure 5. Linear correlations between the adhesion strengths (σ) and the irradiation doses of laser (H) acting on the top surface of the European beech wood samples when using the treatment mode: (**a**) "Laser and PVAc", (**b**) "PVAc and Laser", (**c**) "Laser and PUR", and (**d**) "PUR and Laser".

Figure 6. *Cont.*

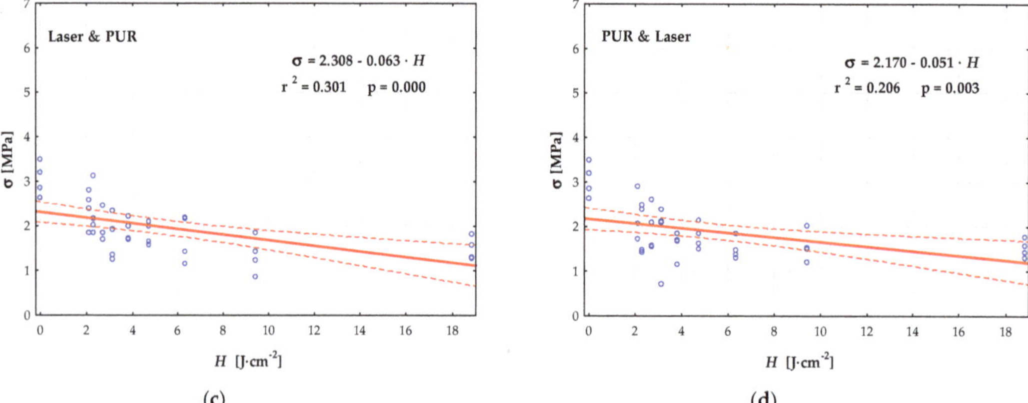

Figure 6. Linear correlations between the adhesion strengths (σ) and the irradiation doses of laser (H) acting on the top surface of the Norway spruce wood samples when using the treatment mode: (**a**) "Laser and PVAc", (**b**) "PVAc and Laser", (**c**) "Laser and PUR", and (**d**) "PUR and Laser".

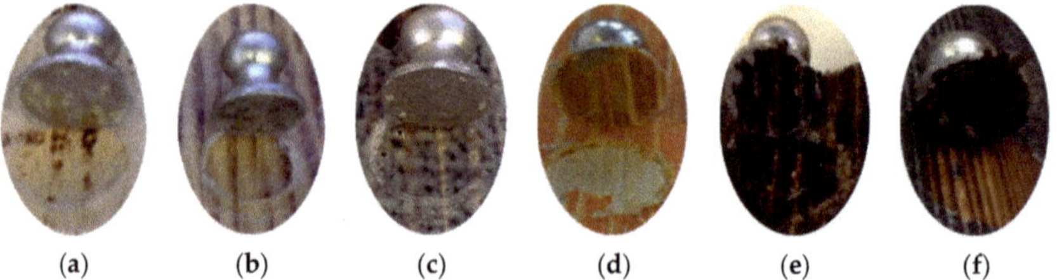

Figure 7. View of the "synthetic polymer–wood" phase interface damage created during the adhesion test of beech samples: (**a**) PVAc and Laser = 3.8 J·cm^{-2}; (**b**) PVAc and Laser = 6.3 J·cm^{-2}; (**c**) PVAc and Laser = 18.8 J·cm^{-2}. Damage created during the adhesion test of spruce samples: (**d**) PVAc and Laser = 3.8 J·cm^{-2}; (**e**) PUR and Laser = 18.8 J·cm^{-2}; (**f**) Laser = 18.8 J·cm^{-2} and PUR. Note: In the adhesion tests, the weakest point for determining the joint was up to 70% of the phase boundary between the wood adherent and the synthetic polymer; the diameter of the steel-roller dolly is 20 mm—i.e., the test area is 3.14 cm^2.

Due to the higher laser irradiation doses (H), in all cases—i.e., for two wood species, two synthetic polymer types, two modes of polymer layer application on the wood surface "before or after laser irradiation"—a significant decrease in the adhesion strength between the synthetic polymer layer and the wood adherent was found. For beech samples, the maximum decrease in the adhesion strength was determined using the highest irradiation dose H of 18.8 J·cm^{-2}. When studying the effect of applying two different synthetic polymers to a beech wood adherent, a milder decrease in the adhesion strength due to an increase in the laser irradiation was found in the application of the PVAc layer (by 43.4% for the mode "Laser and Polymer" or by 47.2% for the mode "Polymer and Laser") than in the application of the PUR layer (by 70.2% for the mode "Laser and Polymer" or by 61.9% for the mode "Polymer and Laser") (Table 3). For spruce samples, the maximum decrease in the adhesion strength was in more cases determined for the application of the second highest irradiation dose H of 9.4 J·cm^{-2} (i.e., by 49.8% or 41.3% when using the PVAc layer, and by 56.1% or 48.5% when using PUR layer for the modes "Laser and Polymer" or "Polymer and Laser") (Table 4).

The above knowledge related to the values of adhesion strength was confirmed by linear correlations ($\sigma = a + b \cdot H$), for which in the beech samples the coefficient of determination r^2 ranged from 0.279 to 0.395 with a level of significance of $p = 0.000$ (Figure 5), while in the spruce samples the r^2 ranged from 0.205 to 0.388 and the p from 0.000 to 0.003 (Figure 6). As mentioned above, for beech samples irradiated with a CO$_2$

laser, the "synthetic polymer–wood" phase interface weakened approximately 1.45 times more when using the PUR polymer layer (by 70.2% or 61.9%) than when using the PVAc one (by 43.4% or 47.2%); this tendency is also documented by the higher negative values of the parameter "b" in the linear correlations when using the PUR layer (Figure 5).

Summarizing the results of all the irradiation modes and doses of a CO_2 laser, the adhesion strength decreased more evidently (1) for the beech versus spruce wood adherent, (2) for the PUR versus PVAc polymer layer, and (3) for the laser beam focusing on the native wood versus synthetic polymer (Figures 5 and 6). This knowledge achieved for the beech wood adherent and the PUR polymer layer can be explained by the greater potential for adhesion strength drop due to the action of a CO_2 laser for such types of wood adherent or polymer layers, which in the initial reference state secured a higher adhesion strength for the "synthetic polymer–wood" phase interface.

3.3. Mold Resistance

The mold growth activity (*MGA*) of the microscopic fungi *Aspergillus niger* and *Penicillium purpurogenum* on the top surfaces of the native wood specimens (i.e., wood surfaces not covered with a layer of synthetic polymer and not modified with a CO_2 laser) was evident from a beginning of the mycological test, whereas on the final 21st day the molds covered more than 30% or 50% of the wood surfaces, with the *MGA* ranging from 3 to 4. This result confirmed the very low mold resistance of the natural/untreated beech [18,55,56] and spruce [57] woods.

The reference wood specimens, meaning specimens covered with a layer of synthetic polymer (i.e., wood surfaces covered with a layer of synthetic polymer but unmodified with a CO_2 laser), were more resistant to microscopic fungi, whereas the *MGA* values ranged from 2 to 3 for the layer with the PVAc polymer or from 0 to 1 for the layer with the PUR polymer (Tables 5 and 6; see Reference).

Table 5. The mold growth activity (*MGA*) evaluated after 21 days on the polymer layer present on the European beech wood adherent. Laser irradiation of specimens with doses (*H*) was performed before or after covering them with a layer of synthetic polymer.

Irradiation Dose H (J·cm^{-2})	Ref. 0	A 2.1	B 2.3	C 2.7	D 3.1	E 3.8	F 4.7	G 6.3	H 9.4	I 18.8	$MGA = f(H)$ r^2; p-Value
Beech				*Aspergillus niger—MGA (0–4)*							
Laser and PVAc	2.67	2.67	2.67	2.33	2.33	2.00	1.67	1.67	2.00	2.33	$MGA = 2.334 - 0.019 \cdot H$ $r^2 = 0.070; p = 0.158$
PVAc and Laser	2.67	3.33	2.67	1.67	1.67	2.00	1.67	1.33	2.00	2.00	$MGA = 2.277 - 0.033 \cdot H$ $r^2 = 0.059; p = 0.196$
Laser and PUR	0.67	0.00	1.00	1.00	1.00	0.67	1.00	0.33	0.00	0.00	$MGA = 0.825 - 0.049 \cdot H$ $r^2 = 0.252; p = 0.005$
PUR and Laser	0.67	1.33	1.00	1.00	0.33	0.33	0.33	0.67	0.67	0.33	$MGA = 0.810 - 0.027 \cdot H$ $r^2 = 0.066; p = 0.170$
Beech				*Penicillium purpurogenum—MGA (0–4)*							
Laser and PVAc	2.33	2.33	2.67	2.33	2.33	2.00	1.33	1.67	1.67	2.00	$MGA = 2.224 - 0.030 \cdot H$ $r^2 = 0.030; p = 0.360$
PVAc and Laser	2.33	2.00	2.33	2.33	1.67	2.00	1.67	1.67	1.33	1.66	$MGA = 2.108 - 0.039 \cdot H$ $r^2 = 0.138; p = 0.043$
Laser and PUR	0.67	1.00	1.00	0.67	1.00	0.67	0.67	0.00	0.00	0.33	$MGA = 0.817 - 0.041 \cdot H$ $r^2 = 0.182; p = 0.019$
PUR and Laser	0.67	1.00	1.00	0.67	0.67	1.00	1.33	1.00	1.00	0.67	$MGA = 0.927 - 0.005 \cdot H$ $r^2 = 0.004; p = 0.733$

Note: Mean value is from 3 replicates. Relation $MGA = f(H)$ is determined from the individual values of 30 specimens. r^2 is the coefficient of determination. p is the level of significance.

Table 6. The mold growth activity (*MGA*) evaluated after 21 days on the polymer layer present on the Norway spruce wood adherent. Laser irradiation of specimens with doses (*H*) was performed before or after covering them with a layer of synthetic polymer.

Irradiation Dose H (J·cm^{-2})	Ref. 0	A 2.1	B 2.3	C 2.7	D 3.1	E 3.8	F 4.7	G 6.3	H 9.4	I 18.8	$MGA = f(H)$ r^2; p-Value
Spruce				*Aspergillus niger—MGA* (0–4)							
Laser and PVAc	2.33	2.00	2.67	2.33	2.33	2.00	1.67	1.33	1.67	1.67	$MGA = 2.334 - 0.019 \cdot H$ $r^2 = 0.157; p = 0.030$
PVAc and Laser	2.33	2.67	2.00	1.67	1.67	2.00	2.00	1.67	1.67	1.00	$MGA = 2.167 - 0.063 \cdot H$ $r^2 = 0.378; p = 0.000$
Laser and PUR	0.67	1.00	0.67	0.33	0.67	0.67	0.67	0.33	0.33	0.00	$MGA = 0.754 - 0.042 \cdot H$ $r^2 = 0.182; p = 0.019$
PUR and Laser	0.67	0.67	0.67	1.00	0.67	0.33	0.33	0.67	0.67	0.67	$MGA = 0.637 - 0.001 \cdot H$ $r^2 = 0.000; p = 0.973$
Spruce				*Penicillium purpurogenum—MGA* (0–4)							
Laser and PVAc	2.33	2.67	2.67	2.00	2.33	2.00	1.67	1.67	2.33	2.00	$MGA = 2.277 - 0.021 \cdot H$ $r^2 = 0.041; p = 0.281$
PVAc and Laser	2.33	2.33	2.00	2.00	2.00	2.33	1.67	2.00	2.00	1.67	$MGA = 2.189 - 0.029 \cdot H$ $r^2 = 0.135; p = 0.046$
Laser and PUR	0.33	0.67	1.00	1.00	0.33	1.00	1.00	0.33	0.67	0.67	$MGA = 0.708 - 0.002 \cdot H$ $r^2 = 0.000; p = 0.927$
PUR and Laser	0.67	0.67	0.67	0.67	0.33	0.33	0.33	1.00	0.67	1.00	$MGA = 0.506 - 0.024 \cdot H$ $r^2 = 0.065; p = 0.175$

Note: Mean value is from 3 replicates. Relation *MGA* = f(*H*) is determined from the individual values of 30 specimens. r^2 is the coefficient of determination. p is the level of significance.

With an increase in the laser irradiation doses (*H*), the *MGA* values of the microscopic fungi determined on the top surfaces of the laser-modified specimens (i.e., the surfaces of specimens modified with a CO_2 laser before or after covering them with a layer of PVAc or PUR polymer) decreased significantly in comparison to the reference specimens only in some cases, as is documented by the results of the linear correlations *MGA* = f(*H*) (Tables 5 and 6).

For the treatment mode "PVAc and Laser", the *MGA* values decreased maximally to 1–2 from 2–3 determined for the reference "PVAc". On the other hand, no significant change in the *MGA* occurred for the specimens treated with the mode "PUR and Laser". This result can be explained by the evidently lower initial mold resistance of the PVAc polymer (Ref. PVAc: *MGA* = 2.33–2.67) in comparison to the very good initial mold resistance of the PUR polymer (Ref. PUR: *MGA* = 0.33–0.67) (Tables 5 and 6).

The *MGA* values usually reduced significantly ($p \leq 0.05$) due to the higher irradiation doses *H* only for specimens whose surfaces were modified with the following two modes: "PVAc and Laser" and "Laser and PUR" (Tables 5 and 6, Figures 8 and 9).

Using the "PVAc and Laser" mode—i.e., when specimens were firstly covered with a layer of PVAc polymer and subsequently irradiated with a CO_2 laser—the linear correlations (*MGA* = a + b · *H*) for the beech specimens had the coefficient of determination r^2 0.059 (*Aspergillus niger*) or 0.138 (*Penicillium purpurogenum*) and the level of significance $p = 0.196$ (nonsignificant) or 0.043 (significant at the 95% level). The linear correlations for the spruce specimens were partly more representative, with the r^2 0.378 or 0.135 and the $p = 0.000$ (significant at the 99.9% level) or 0.046 (significant at the 95% level) (Tables 5 and 6, Figures 8 and 9).

Using the treatment mode "Laser and PUR", a significant decrease in mold activity was found, with the *MGA* values of up to 0. In the linear correlations (*MGA* = a + b · *H*) for beech specimens had the coefficient of determination r^2 0.252 (*Aspergillus niger*) or 0.182 (*Penicillium purpurogenum*) and the level of significance $p = 0.005$ or 0.019 (Table 4), while spruce specimens had r^2 0.182 or 0.000 and $p = 0.019$ or 0.927 (Table 5). These tendencies toward the total stopping of mold activity on the PUR polymer layer present on the laser-irradiated wood adherent could be explained by the creation of mold-inhibitory substances during the laser modification of wood surface with their ability to penetrate through a liquid not-yet-hardened layer of PUR polymer before its curing on the top surface of specimen (i.e., on the top layer of the PUR polymer) exposed to the mycological test with molds. However, in the future this hypothesis has to be confirmed by physical and chemical analyses.

Beech

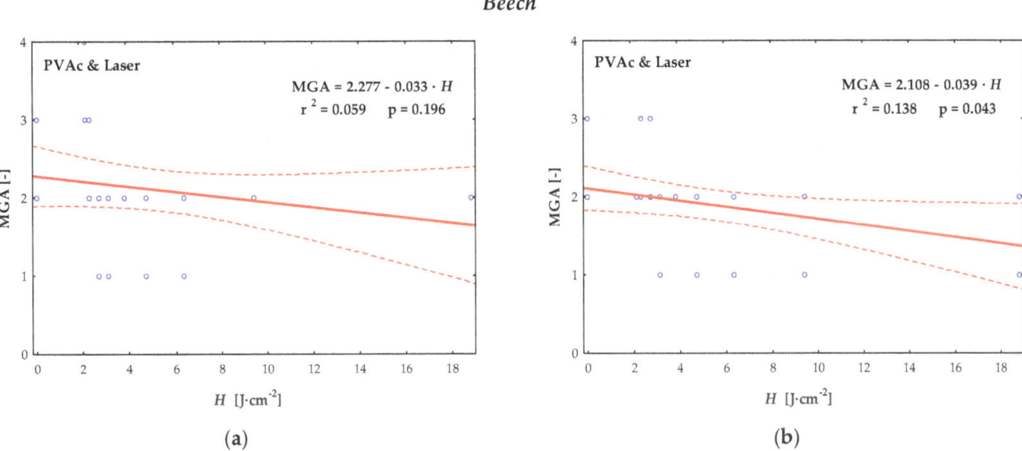

Figure 8. Linear correlations between the mold growth activities (*MGA*) of *Aspergillus niger* (**a**) or *Penicillium purpurogenum* (**b**) and the laser irradiation doses (*H*) acting on the top surface of the European beech wood samples when using the treatment mode "PVAc and Laser".

Spruce

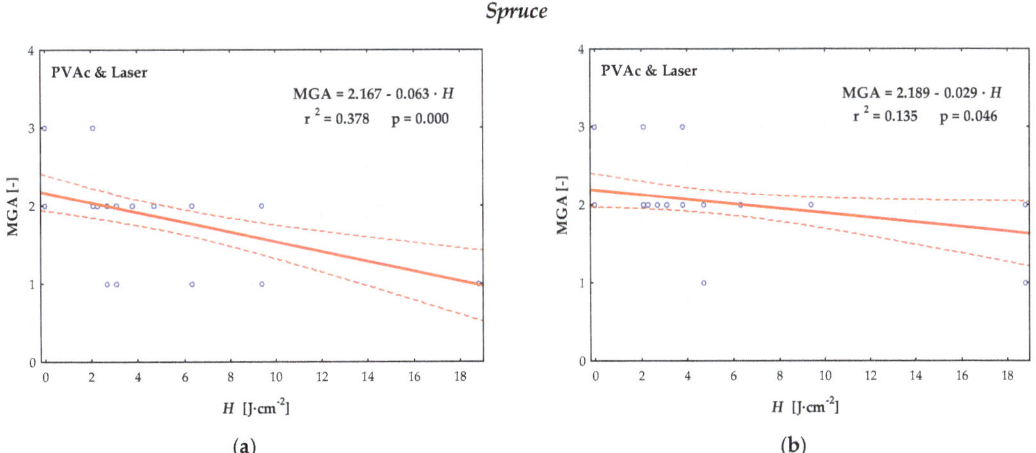

Figure 9. Linear correlations between the mold growth activities (*MGA*) of *Aspergillus niger* (**a**) or *Penicillium purpurogenum* (**b**) and the laser irradiation doses (*H*) acting on the top surface of the Norway spruce wood samples when using the treatment mode "PVAc and Laser".

4. Conclusions

The increased doses H of CO_2 laser from 2.1 to 18.8 $J\cdot cm^{-2}$ acting on the native beech and spruce wood surfaces or on the layers of synthetic PVAc and PUR polymers present on wood surfaces caused the decomposition and carbonization of wood components or synthetic polymers—visible by their darkening until blacking and the creation of air bubbles in synthetic polymers.

The adhesion strength between the synthetic polymer layer and the wood adherent decreased continuously in all cases with the increase in laser doses H from 2.1 $J\cdot cm^{-2}$ until 18.8 $J\cdot cm^{-2}$. The decrease in the adhesion strength when using H of 18.8 $J\cdot cm^{-2}$ was from 41.3% (spruce: "PVAc and Laser") up to 70.2% (beech: "Laser and PUR"). Due to the laser irradiation, the adhesion strength decreased more evidently for the beech wood versus the spruce wood adherent, for the PUR polymer versus the PVAc polymer layer, and for the laser beam focusing on the native wood versus the synthetic polymer.

The mold growth activities of the microscopic fungi *Aspergillus niger* and *Penicillium purpurogenum* on the surfaces of the tested specimens were evidently inhibited by the CO_2 laser only if the mode "PVAc and Laser" was used—i.e., if the wood specimens were firstly covered with a less mold-resistant PVAc polymer and subsequently irradiated with laser doses $H \geq 3.1\ J\cdot cm^{-2}$.

The achieved results indicated that the studied laser technological operations—(1) the laser pretreatment of wood surfaces before covering them with PVAc or PUR polymers, or (2) the laser treatment of PVAc or PUR layers present on the wood surface—are not best suited for practical use. In this light, laser-machined wood should not be directly bonded or painted with polymer adhesives or coatings and laser beams should not be focused on wood surfaces covered with synthetic polymer layers.

Author Contributions: Conceptualization, L.R. and Z.V.; methodology, L.R. and Z.V.; software, L.R. and Z.V.; validation, L.R. and Z.V.; formal analysis, L.R. and Z.V.; investigation, L.R. and Z.V.; resources, L.R. and Z.V.; data curation, L.R. and Z.V.; writing—original draft preparation, L.R. and Z.V.; writing—review and editing, L.R. and Z.V.; visualization, L.R. and Z.V.; supervision, L.R. and Z.V.; project administration, L.R.; funding acquisition, L.R. All authors have read and agreed to the published version of the manuscript.

Funding: This work was supported by the Slovak Research and Development Agency under the contract no. APVV-17-0583 and the VEGA project 1/0729/18.

Institutional Review Board Statement: Not applicable.

Informed Consent Statement: Not applicable.

Data Availability Statement: Data sharing is not applicable to this article.

Acknowledgments: The authors would like to thank the Slovak Research and Development Agency under contract no. APVV-17-0583 and also the VEGA project 1/0729/18 for the funding and financial support. This publication is also the result of the following project implementation: Progressive research of performance properties of wood-based materials and products (LignoPro), ITMS 313011T720 supported by the Operational Program Integrated Infrastructure (OPII) funded by the ERDF. For help with the experiments, the authors thank also Ing. Jozef Valko and Ing. Matúš Nemeček.

Conflicts of Interest: The authors declare no conflict of interest.

References

1. Acda, M.N.; Devera, E.E.; Cabangon, R.J.; Ramos, H.J. Effects of plasma modification on adhesion properties of wood. *Int. J. Adhes. Adhes.* **2011**, *32*, 70–75. [CrossRef]
2. Petrič, M. Surface Modification of Wood. *Rev. Adhes. Adhes.* **2013**, *1*, 216–247. [CrossRef]
3. Gurau, L.; Petru, A.; Varodi, A.; Timar, M.C. The Influence of CO_2 Laser beam power output and scanning speed on surface roughness and colour changes of beech (*Fagus sylvatica*). *BioResources* **2017**, *12*, 7395–7412. [CrossRef]
4. Nath, S.; Waugh, D.G.; Ormondroyd, G.A.; Spear, M.J.; Pitman, A.J.; Sahoo, S.; Curling, S.F.; Mason, P. CO_2 laser interactions with wood tissues during single pulse laser-incision. *Opt. Laser Technol.* **2020**, *126*, 106069. [CrossRef]

5.   Petutschnigg, A.; Stöckler, M.; Steinwendner, F.; Schnepps, J.; Gütler, H.; Blinzer, J.; Holzer, H.; Schnabel, T. Laser Treatment of Wood Surfaces for Ski Cores: An Experimental Parameter Study. *Adv. Mater. Sci. Eng.* **2013**, *2013*, 1–7. [CrossRef]

6.   Kúdela, J.; Kubovský, I.; Andrejko, M. Surface Properties of Beech Wood after CO_2 Laser Engraving. *Coatings* **2020**, *10*, 77. [CrossRef]

7.   Iždinský, J.; Reinprecht, L.; Sedliačik, J.; Kúdela, J.; Kučerová, V. Bonding of Selected Hardwoods with PVAc Adhesive. *Appl. Sci.* **2020**, *11*, 67. [CrossRef]

8.   Slabejová, G.; Langová, N.; Deáková, V. Influence of silicone resin modification on veneer tensile strength and deformation. *Acta Fac. Xylologiae Zvolen* **2017**, *59*, 41–47. [CrossRef]

9.   Gaff, M.; Razaei, F.; Sikora, A.; Hysek, Š.; Sedlecký, M.; DiTommaso, G.; Corleto, R.; Kamboj, G.; Sethy, A.; Vališ, M.; et al. Interactions of monitored factors upon tensile glue shear strength on laser cut wood. *Compos. Struct.* **2020**, *234*, 111679. [CrossRef]

10.   Prayitno, T.A.; Widyorini, R.; Lukmandaru, G. The adhesion properties of wood preserved with natural preservatives. *Wood Res.* **2016**, *61*, 197–204.

11.   Jablonský, M.; Šmatko, L.; Botková, M.; Tiňo, R.; Šima, J. Modification of wood wettability (European Beech) by diffuse coplanar surface barrier discharge plasma. *Cellul. Chem. Technol.* **2014**, *50*, 41–48.

12.   Žigon, J.; Petrič, M.; Dahle, S. Dielectric barrier discharge (DBD) plasma pretreatment of lignocellulosic materials in air at atmospheric pressure for their improved wettability: A literature review. *Holzforschung* **2018**, *72*, 979–991. [CrossRef]

13.   Reinprecht, L.; Tiňo, R.; Šomšák, M. The Impact of Fungicides, Plasma, UV-Additives and Weathering on the Adhesion Strength of Acrylic and Alkyd Coatings to the Norway Spruce Wood. *Coatings* **2020**, *10*, 1111. [CrossRef]

14.   Wascher, R.; Bittner, F.; Avramidis, G.; Bellmann, M.; Endres, H.-J.; Militz, H.; Viöl, W. Use of computed tomography to determine penetration paths and the distribution of melamine resin in thermally-modified beech veneers after plasma treatment. *Compos. Part A Appl. Sci. Manuf.* **2020**, *132*, 105821. [CrossRef]

15.   Barcikowski, S.; Koch, G.; Odermatt, J. Characterisation and modification of the heat affected zone during laser material processing of wood and wood composites. *Holz Roh. Werkst.* **2006**, *64*, 94–103. [CrossRef]

16.   Islam, N.; Ando, K.; Yamauchi, H.; Kobayashi, Y.; Hattori, N. Passive impregnation of liquid in impermeable lumber incised by laser. *J. Wood Sci.* **2007**, *53*, 436–441. [CrossRef]

17.   Kubovský, I.; Kačík, F.; Reinprecht, L. The impact of UV radiation on the change of colour and composition of the surface of lime wood treated with a CO_2 laser. *J. Photochem. Photobiol. A Chem.* **2016**, *322*, 60–66. [CrossRef]

18.   Vidholdová, Z.; Reinprecht, L.; Igaz, R. The Impact of Laser Surface Modification of Beech Wood on its Color and Occurrence of Molds. *BioResources* **2017**, *12*, 4177–4186. [CrossRef]

19.   Chernykh, M.; Kargashina, E.; Stollmann, V. The use of wood veneer for laser engraving production. *Acta Fac. Xylologiae Zvolen* **2018**, *60*, 121–127. [CrossRef]

20.   Mertens, N.; Wolkenhauer, A.; Leck, M.; Viöl, W. UV laser ablation and plasma treatment of wooden surfaces—A comparing investigation. *Laser Phys. Lett.* **2006**, *3*, 380–384. [CrossRef]

21.   Novák, I.; Chodák, I.; Sedliačik, J.; Vanko, V.; Matyašovský, J.; Šivová, M. Pre-treatment of beech wood by cold plasma. *SGGW For. Wood Technol.* **2013**, *83*, 288–291.

22.   Král, P.; Ráhel', J.; Stupavská, M.; Šrajer, J.; Klímek, P.; Mishra, P.K.; Wimmer, R. XPS depth profile of plasma-activated surface of beech wood (*Fagus sylvatica*) and its impact on polyvinyl acetate tensile shear bond strength. *Wood Sci. Technol.* **2015**, *49*, 319–330. [CrossRef]

23.   Reinprecht, L. Transport of preservatives into wood. In *Better Wood Products through Science*; IUFRO Division 5: Nancy, France, 1992; p. 453.

24.   Aligizaki, E.M.; Melessanaki, K.; Pournou, A. The use of lasers for the removal of shellac from wood. *e-Preserv. Sci.* **2008**, *5*, 36–40.

25.   Wang, Y.; Ando, K.; Hattori, N. Changes in the anatomy of surface and liquid uptake of wood after laser incising. *Wood Sci. Technol.* **2012**, *47*, 447–455. [CrossRef]

26.   Fukuta, S.; Nomura, M.; Ikeda, T.; Yoshizawa, M.; Yamasaki, M.; Sasaki, Y. UV laser machining of wood. *Holz Roh. Werkst.* **2016**, *74*, 261–267. [CrossRef]

27.   Martínez-Conde, A.; Krenke, T.; Frybort, S.; Müller, U. Review: Comparative analysis of CO_2 laser and conventional sawing for cutting of lumber and wood-based materials. *Wood Sci. Technol.* **2017**, *51*, 943–966. [CrossRef]

28.   Kúdela, J.; Reinprecht, L.; Vidholdová, Z.; Andrejko, M. Surface properties of beech wood modified by a CO_2 laser. *Acta Fac. Xylologia Zvolen* **2019**, *61*, 5–18. [CrossRef]

29.   Jiang, T.; Yang, C.; Yu, Y.; Doumbia, B.S.; Liu, J.; Ma, Y. Prediction and Analysis of Surface Quality of Northeast China Ash Wood during Water-Jet Assisted CO_2 Laser Cutting. *J. Renew. Mater.* **2021**, *9*, 119–128. [CrossRef]

30.   Yung, K.C.; Choy, H.S.; Xiao, T.; Cai, Z. UV laser cutting of beech plywood. *Int. J. Adv. Manuf. Technol.* **2021**, *112*, 925–947. [CrossRef]

31.   Kačík, F.; Kubovský, I. Chemical changes of beech wood due to CO_2 laser irradiation. *J. Photochem. Photobiol. A Chem.* **2011**, *222*, 105–110. [CrossRef]

32.   Panzner, M.; Wiedemann, G.; Henneberg, K.; Fischer, R.; Wittke, T.; Dietsch, R. Experimental investigation of the laser ablation process on wood surfaces. *Appl. Surf. Sci.* **1998**, *127*, 787–792. [CrossRef]

33.   Dolan, J.A. Characterization of Laser Modified Surfaces for Wood Adhesion. Master's Thesis, Macromolecular Science and Engineering, Faculty of Virginia Polytechnic Institute, Blacksburg, VA, USA, 2014; 100p.

34. Aniszewska, M.; Maciak, A.; Zychowicz, W.; Zowczak, W.; Mühlke, T.; Christoph, B.; Lamrini, S.; Sujecki, S. Infrared Laser Application to Wood Cutting. *Materials* **2020**, *13*, 5222. [CrossRef] [PubMed]

35. Li, R.; Xu, W.; Wang, X.; Wang, C. Modeling and predicting of the color changes of wood surface during CO_2 laser modification. *J. Clean. Prod.* **2018**, *183*, 818–823. [CrossRef]

36. Açık, C.; Tutuş, A. The effect of traditional and laser cutting on surface roughness of wood materials used in furniture industry. *Wood Ind. Eng.* **2020**, *2*, 45–50.

37. Dolan, J.A.; Sathitsuksanoh, N.; Renneckar, S.; Rodriguez, K.; Simmons, B.A.; Frazier, C.E. Biocomposite adhesion without added resin: Understanding the chemistry of the direct conversion of wood into adhesives. *RSC Adv.* **2015**, *5*, 67267–67276. [CrossRef]

38. Ozdemir, M.; Sadikoglu, H. A new and emerging technology: Laser-induced surface modification of polymers. *Trends Food Sci. Technol.* **1998**, *9*, 159–167. [CrossRef]

39. Ülker, O. Wood adhesives and bonding theory. In *Adhesives–Application and Properties*, 1st ed.; Rudawska, A., Ed.; IntechOpen: London, UK, 2016; pp. 271–288.

40. Reinprecht, L. *Wood Deterioration, Protection and Maintenance*; John Wiley & Sons, Ltd.: Chichester, UK, 2016; 357p.

41. Imken, A.A.; Brischke, C.; Kögel, S.; Krause, K.C.; Mai, C. Resistance of different wood-based materials against mould fungi: A comparison of methods. *Holz Roh. Werkst.* **2020**, *78*, 661–671. [CrossRef]

42. Jeřábková, E.; Tesařová, D.; Polášková, H. Resistance of various materials and coatings used in wood constructions to growth of microorganisms. *Wood Res.* **2018**, *63*, 993–1002.

43. Vidholdová, Z.; Slabejová, G. Environmental valuation of selected transparent wood coatings from the view of fungal resistance. *SGGW For. Wood Technol.* **2018**, *103*, 164–168.

44. Frihart, C.R. Wood adhesion and adhesives. In *Handbook of Wood Chemistry and Wood Composites*, 2nd ed.; Rowell, R.M., Ed.; CRC Press Taylor & Francis Group: Boca Raton, FL, USA, 2013; pp. 255–319.

45. Greń, I.; Gąszczak, A.; Guzik, U.; Bartelmus, G.; Łabużek, S. A comparative study of biodegradation of vinyl acetate by environmental strains. *Ann. Microbiol.* **2011**, *61*, 257–265. [CrossRef] [PubMed]

46. Howard, G.T. Biodegradation of polyurethane: A review. *Int. Biodeterior. Biodegrad.* **2002**, *49*, 245–252. [CrossRef]

47. EN ISO Paints and Varnishes. *Pull-Off Test for Adhesion*; European Committee for Standardization: Brussels, Belgium, 2016.

48. EN Paints and Varnishes. *Laboratory Method for Testing the Efficacy of Film Preservatives in a Coating against Fungi*; European Committee for Standardization: Brussels, Belgium, 2014.

49. Camino, G.; Lomakin, S. Intumescent materials. In *Fire Retardant Materials*, 1st ed.; Horrocks, A.R., Price, D., Price, D., Eds.; Woodhead Publishing Ltd.: Cambridge, UK, 2001; pp. 318–336.

50. Dasari, A.; Yu, Z.-Z.; Cai, G.-P.; Mai, Y.-W. Recent developments in the fire retardancy of polymeric materials. *Prog. Polym. Sci.* **2013**, *38*, 1357–1387. [CrossRef]

51. Rodriguez-Vazquez, M.; Liauw, C.M.; Allen, N.S.; Edge, M.; Fontan, E. Degradation and stabilisation of poly(ethylene-stat-vinyl acetate): 1-Spectroscopic and rheological examination of thermal and thermo-oxidative degradation mechanisms. *Polym. Degrad. Stab.* **2006**, *91*, 154–164. [CrossRef]

52. Rimez, B.; Rahier, H.; Van Assche, G.; Artoos, T.; Biesemans, M.; Van Mele, B. The thermal degradation of poly(vinyl acetate) and poly(ethylene-co-vinyl acetate), Part I: Experimental study of the degradation mechanism. *Polym. Degrad. Stab.* **2008**, *93*, 800–810. [CrossRef]

53. Richter, K.; Pizzi, A.; Despres, A. Thermal stability of structural one-component polyurethane adhesives for wood—structure-property relationship. *J. Appl. Polym. Sci.* **2006**, *102*, 5698–5707. [CrossRef]

54. Golling, F.E.; Pires, R.; Hecking, A.; Weikard, J.; Richter, F.; Danielmeier, K.; Dijkstra, D. Polyurethanes for coatings and adhesives—Chemistry and applications. *Polym. Int.* **2019**, *68*, 848–855. [CrossRef]

55. Hamed, S.A.M.; Mansour, M.M.A. Comparative study on micromorphological changes in wood due to soft-rot fungi and surface mold. *Sci. Cult.* **2018**, *4*, 35–41. [CrossRef]

56. Reinprecht, L.; Vidholdová, Z.; Iždinský, J. Bacterial and mold resistance of selected tropical wood species. *BioResources* **2020**, *15*, 5198–5209. [CrossRef]

57. Kumar, A.; Ryparová, P.; Škapin, A.S.; Humar, M.; Pavlič, M.; Tywoniak, J.; Hajek, P.; Žigon, J.; Petrič, M. Influence of surface modification of wood with octadecyltrichlorosilane on its dimensional stability and resistance against Coniophora puteana and molds. *Cellulose* **2016**, *23*, 3249–3263. [CrossRef]

 *forests*

Review

Expandable Graphite as a Fire Retardant for Cellulosic Materials—A Review

Bartłomiej Mazela *, Anyelkis Batista and Wojciech Grześkowiak

Faculty of Wood Technology, Poznan University of Life Sciences, ul. Wojska Polskiego 28,
60-637 Poznan, Poland; anielkis.batista@up.poznan.pl (A.B.); wojciech.grzeskowiak@up.poznan.pl (W.G.)
* Correspondence: bartlomiej.mazela@up.poznan.pl

Received: 22 May 2020; Accepted: 7 July 2020; Published: 13 July 2020

Abstract: A diversity of chemicals is used to produce fire retardants (FRs); some of the main group of chemicals are hazardous to the environment as well as to human life; however, expandable graphite (EG) can be a gateway to a more environmentally friendly FRs or intumescent fire retardants (IFRs). Researchers define intumescent as the swelling of a particular substance placed between a heat source and an underlying substrate when they are heated. EG is a material with extraordinary thermophysical and mechanical properties. The referred EG properties are unparalleled. EG is a low-density carbon material having a series of unique properties: developed specific surface, binder-free pressing capacity, stability to aggressive media, and low thermal conductivity. Therefore, EG is a promising material both for research work and for industrial applications. The primary goal of this literature review was to report current knowledge on the use of EG as a fire retardant for cellulose and cellulose-modified materials. EG is produced, among other methods, by thermal shock of graphite oxide under forming gas. When exposed to heat, EG will expand. The expansion mechanism was presented in this review. Equally important to this review is the knowledge related to cellulose thermal degradation and cellulose impact on the development of science and technology.

Keywords: cellulose; modification; coating; expandable graphite; fire retardant; thermal degradation

---

1. Introduction to the Mechanism of Fire Retardants

Fire retardants (FRs) are chemicals added to combustible materials to improve their resistance to ignition [1]. Depending on the processing methods, fire retardants (FR) are often categorized as an additive or reactive compounds. Additive FRs are often blended into the polymer matrix during its processing. These FRs do not chemically react with the polymer. Reactive FRs are dose polymerized with resin during the processing of cellulosic material to become integrated into its molecular network structure. These FRs are also known as chemically-modified FRs [2]. The footprints of fire retardants point to an effective ideal fire retardant that should comprise of the following characteristics: thermal stability, compatibility with the protected polymer, no change in the physical-chemical properties of the protected polymer, and low toxicity under heat exposure or during burning [3].

The amount of flame retardant that must be added to achieve the desired level of fire safety can range from less than 1% for highly effective flame retardants up to more than 50% for inorganic fillers; the typical range is 5% to 20%, by weight [1,4].

In works of Camino and Costa [3] and Nazaré [5], two types of FR mechanism can be identified: gas-phase mechanism and/or condensed phase mechanism. Whatever the case, the main goal of FR systems is to reduce the heat released from the polymer under combustion to levels that lead to flame stability of the polymer. This can be achieved by modifying the rate of chemical or physical processes taking place in one or more of the steps of the burning mechanism of the material [6].

In order to understand the mechanisms of fire retardants, it is required to have detailed knowledge of the mechanism of the thermal degradation process of the polymer and the thermal correlation of the additive and of the polymer-additive mixture. Organic polymeric materials (natural or synthetic) may start and/or propagate fire (thermal degradation) due to heat exposure involving combustible products in various concentrations over different temperature ranges [6,7].

In general, and according to studies [6], three main processes are involved in the thermal degradation of the majority of thermoplastic polymers. Focusing on the molecular weight (MW) of the polymer during the thermal degradation process, most authors agree that on the first stage, there is a very slight reduction in MW, on the middle stage, there is a slow decrease in MW, and on the final stage, a rapid drop in MW occurs.

These FRs are designed to minimize the risk of a fire starting in case of contact with a small heat source, such as a cigarette, candle, or an electrical fault. As organic and/or inorganic formulation, FR based on inorganic formulation may include aluminum trioxide ($Al(OH)_3$), magnesium hydroxide ($Mg(OH)_2$), Antimony trioxide (Sb_2O_3), ammonium polyphosphate ($(NH_4PO_3)_n$), red phosphorus, boric acid (H_3BO_3), and others. This group represents about 50% by volume of the global FR production. The remaining categories are organic FRs [1,8]. FRs are used in buildings and constructions, electronics, automotive and transportation, wires, cables, textiles, and others.

Environmental problems, toxicity, and hazards to human life are major factors that will force the reduction of use or even the total abandonment of these chemicals. Most fire retardant formulations are halogen-based. Inorganic acids, nitrogen, phosphorous, and intumescent-based FR formulations have been and will continue to be important in flame retardant researches. However, environmental concerns and health care issues are stimulating a strong interest in innovative alternatives to fire retardant systems.

Intumescent flame retardants are the most promising candidate to substitute the halogen-containing flame retardants, which are free of halogen and with relatively high FR efficiency. In general, the intumescent fire retardant (IFR) systems are composed of three components, i.e., an acid source, a carbonization agent (or char forming agent), and a blowing agent. However, the traditional IFR additives are susceptible to migration onto the polymer surface during the processing due to their low molecular weight and thus decrease the FR efficiency. To solve the shortcomings, high molecular weight, namely, oligomeric or polymeric, IFRs have been developed, which provide a good strategy to solve the above problems [9].

In general, an IFR mechanism works on the basis of promoting a char layer between the heat source and the protected substrate during expansion due to heat exposure [10,11]. A char layer that might be able to protect the underlying structure should simultaneously combine the following characteristics: low thermal conductivity in the thickness direction, high in-planar thermal, high stability, low heat absorption, high heat capacity, and a compact structure (low porosity and permeability system) to limit mass loss in thickness direction fundamentally [12]. Other advantages of using IFR are decreased flame spread, low smoke emission, and the performance of the substrate would not be altered. Applying IFR coatings is one of the most efficient ways of providing fire retardancy to flammable materials [11].

2. Combustion Properties of Cellulose

Cellulose is present in large volumes on the planet. As one of the most abundant materials on earth, cellulose is the most common organic polymer [13], representing more than 1 trillion tons of annual biomass production. Cellulose is considered the main chemical component in many lignocellulosic resources. Of all biomass, cellulose (Figure 1) represents more than 50% by weight. It is comprised of two anhydroglucose rings ($(C_6H_{10}O_5)_n$) with a linear homopolymer of glucopyranose residues connected by β-1, 4-glycosidic bond. Understanding the pyrolysis of cellulose can be beneficial towards a successful explanation of its pyrolytic mechanism [14]. The degree of polymerization, n, varies between 10,000 and 15,000, where n is dependent on the cellulose source material. A cellulose molecule unit is presented in Figure 1.

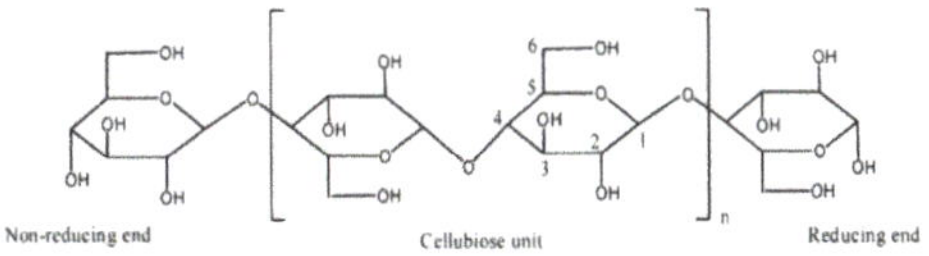

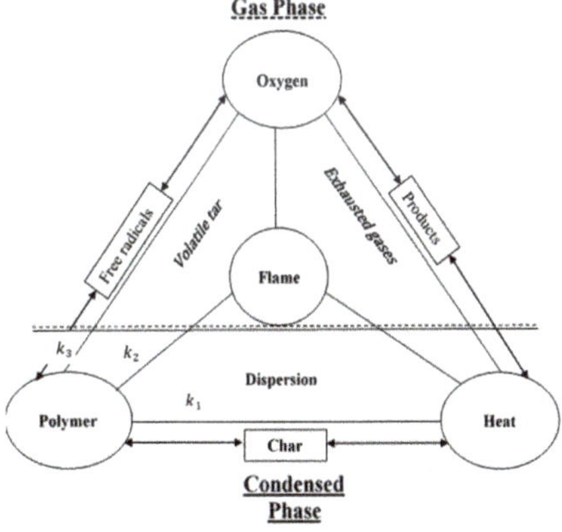

Figure 1. Cellulose molecule unit [14].

2.1. Combustibility of Cellulose

Cellulose derived from wood pulp has an average of 3000 repeated units. A large number of hydroxyl groups on the sugar molecule, which leaves the polymer as water molecules during the decomposition, results in char formation [15]. Due to its strength, charring ability, and biodegradability, cellulose-based reinforced polymers have received considerable attention. The burning characteristics of cellulosic fibers vary depending on their chemical composition. For example, lignin, as natural polymer, could improve the flame resistance of the fibers; in fact, it is reported that a low concentration of lignin (~10 wt%) in cellulose fiber improves the char formation during thermal decomposition processes [16].

2.2. Thermal Decomposition of Cellulose

During the decomposition (combustion) of flammable materials, some carbonaceous substances tend to swell. Some methods of anti-fire protection achieve flame retarding action by accelerating the formation of a char layer, derived from the intumescent carbon located on the surface of the treated cellulosic material [17]. The comprehension of the pyrolytic behavior of cellulose is essential to biomass thermochemical conversion. Early work for cellulose pyrolysis emphasizes the classic kinetic schemes of three main chemical pathways. It has been found that at low temperatures, the initial process is delayed, which leads to a reduction in the degree of polymerization and the formation of active cellulose [18]. On the other hand, high-temperature pyrolysis of cellulose is developed through two competitive degradation reactions: the first reaction is fundamentally the formation of char and gas. In the second reaction, the degradation process promotes the formation, mainly of tars. Recent researches have excluded the concept of anhydrocellulose formation during the degradation process [18].

Schematically, the literature shows a diversity of diagrams that usually emphasize three main points on cellulose thermal degradation (the polymer, char formation and volatile tars, char and gas). These three points are related to three kinetic points (k_1, k_2, and k_3) [14]. A combination of all these schematics in a two-phase diagram (condensed and gas phase) is presented in Figure 2.

Figure 2. Concept of the combustion Emmons triangle modification [19]; kinetic points (k_1, k_2, k_3).

Although cellulose is often overlooked as a flame retardant for polymers due to its low thermal stability, its ability to charring can be beneficial in reducing calorific value as fuel and reducing peak heat release in the event of a real fire [20]. No report has been found on the concept of "flame volume" regarding cellulose thermal decomposition; some articles present the concept of ignition of solid (cellulosic) fuel and flame spread independently; however, there is no revelation of "flame volume". The concept of "flame volume" based on the volume of cellulose exposed to heat and to flame, consequently, is important to fire safety. Nonetheless, the primary goal of this literature review was to investigate the promising pathways and methods of improving the flammability properties of Cellulosic Material (CellMat)–with the use of expandable graphite (EG).

Recent investigation on intumescent base fire retardants points that it is efficient to create a physical barrier between the possible exposed material and the flame. The development of EG-based flame retardants for several different materials is the new trend in distinguished research projects.

3. Expandable Graphite (EG)

EG is first reported in the literature in 1841 by Schafhautl while analyzing crystal flakes of graphite in a solution of sulfuric acid [21]. The flame retardant performance of EG in a polyurethane coating has been previously reported. The oxygen index increases from 22 to 42 vol.% at 25 wt.% loading. It has been demonstrated that EG is an efficient additive, which acts as a blowing agent as well as a carbonization agent. EG is an intumescent additive known to improve fire retardancy properties in various materials and, in particular, in polyurethane (PU) foam [20]. EG is achieved through flake graphite intercalation derived when flake graphite is exposed to concentrated sulfuric acid (H_2SO_4) in combination with other strong oxidizers, such as nitric acid (NH_3) or potassium permanganate ($KMnO_4$). Flake graphene is synthesized from graphite or other sources of carbon by a top-down method [22]. The oxidation of graphite in protonated solvents leads to graphite oxide, which consists of multiple layers of graphene oxide, with hexagonal carbon structure, which includes hydroxyl group, alkoxy, carbonyl, and acids group. At the industrial scale, the production methods of EG is similar to that of graphite oxide (GO). Due to its economic and environmental advantages, this novel product is attracting attention in many areas of research. However, none of the reviewed papers has assessed the environmental and health hazards of expandable graphite. EG is only presented as a novel and promising alternative flame retardant. EG is not classified according to European Countries (EC) regulations, and no data except the Chinese market about effects on health and environment have been found in the literature [23]. The raw material (graphite) is not classified according to EC regulations. Data about health aspects indicates that there is a risk of physical effects, such as dust bronchitis, in workers engaged in the graphite industry. Sulfuric acid, for example, is a strong acid and can cause severe burns when used in a concentrated form. None of these sparse data indicates any negative health and environmental effects from the use of expandable graphite as an FR component in polyolefin plastics [24]. EG is one of the materials at the forefront of environmentally friendly novels. As an excellent platform to remove oil from the aqueous medium, EG/manganese ferrite composite is used to remove heavy oil from the aqueous medium, and it can be used in oil spills catastrophe as a form of crude oil removal material [25].

Considering graphite, a relatively stable substance at environmental temperature, their expandable property is significantly important to fire retardants' formulation design. When EG is used as flame retardant, its dilatability and thermal stability are very important parameters. In terms of thermostability, EG can be divided into three kinds: low (between 80 and 150 °C), middle (between 180 and 240 °C), and high (between 250 and 300 °C) [26]. At temperatures ranging between 280 and 438 °C, high thermostable graphite will expand, creating a porous physical barrier between the material to be protected and the flame [6]. Moreover, during the char formation, void spaces are formed within char, allowing airflow, and this cool down the fire environment (atmosphere), increasing the time to ignition of protected cellulosic material. In general, the kinetics of fire-protected cellulosic materials are still under investigation [27]. The effectiveness of these flame retardants depends on the heat-induced decomposition of the organic components and the creation of a char layer that insulates the substrate from the heat source. However, as intumescent is required to address more severe and diverse applications, new approaches are needed that provide improved performance over conventional systems. EG flake is a different intumescent additive. As a blowing agent, EG expands up to 100 times its original thickness [4]; it has an isolative layer larger in thickness more than many intumescents. Unlike the carbon char layer formed with chemical intumescent, the graphite-based char formed from expandable flake retains the superiority heat resistance in comparison with other IFRs. In addition, graphite flake is the only intumescent that expands with sufficient force to allow its use in rigidized systems, such as those employing cured phenolic resins [28,29].

4. Expansion Mechanism of EG

The actual cause of expansion/exfoliation is the increase in volume and resultant pressure, caused by the rapid heating of the intercalant. A simplified way to view the process is to model the intercalant as a liquid or solid phase that is fixed between the graphene layers. Heating of the treated graphite results in the conversion of the intercalant from a liquid or solid phase to a gas phase. Gas formation results in an increase in the volume of the intercalant by approximately 100 times [30]. The pressure generated by this volume increase forces the adjacent graphite layers to separate. Figure 3 shows a schematic of the expansion agent in their metastable residence between adjacent graphene layers. The interlayer dimensions are equal to approximately 3.355 Å (6.7 Å/2). The interlayer spacing, post exfoliation, is some value greater than 3.35 Å [21]. An increase in graphene layer spacing is a result of the protection process.

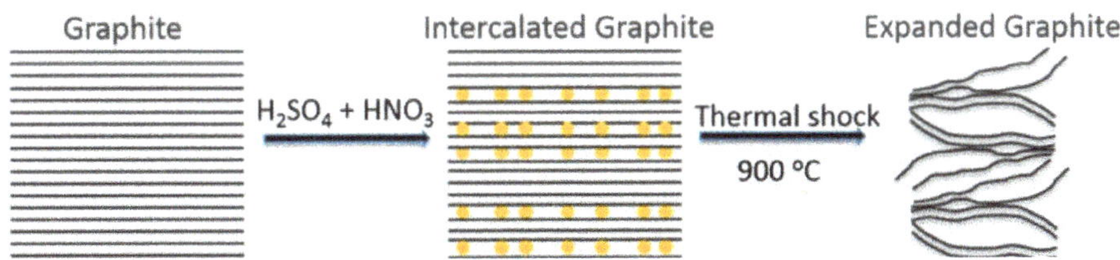

Figure 3. Schematic of expandable graphite (EG) expansion [31].

5. EG as Fire Retardant Additive in Different Materials

The FR performance of EG in a PU coating has been previously reported, where oxygen index increases from 22 to 42 vol.% at 25 wt.% loading. It has been demonstrated that EG is an efficient additive, which acts as a blowing agent as well as a carbonization agent [20].

In the work of Camino et al. [8], thermogravimetric analysis of the EG was carried out through pyrolysis. The unit raising the temperature was 10 °C/min, and, step by step, the density of residue was measured.

Similar tests were performed by Zheng et al. [27], with a temperature increase of 5 °C/min. The course of EG thermal decomposition is represented by the thermogravimetric (TG) (Figure 4) and derivative thermogravimetric (DTG) (Figure 5) curves.

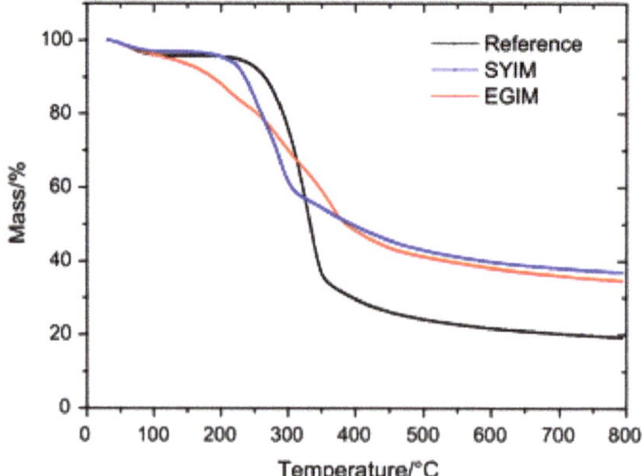

Figure 4. TG (thermogravimetric) curves of synergetic fire retardants insulating material (SYIM) and expandable graphite insulating material (EGIM) [27].

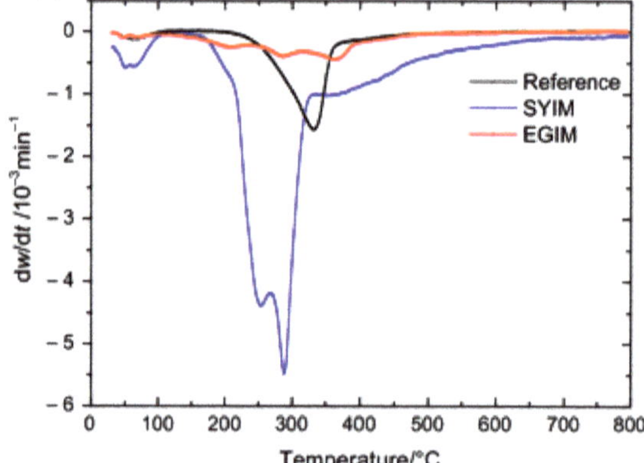

Figure 5. Derivative curves of synergetic fire retardants insulating material (SYIM) and expandable graphite insulating material (EGIM) [27].

In this experiment, the thermobalance could not be used because of the EG expansion, which was observed to expand ten times its original sample size. EG stability was lower at temperatures ranging between 200 and 350 °C, with a maximum rate of weight loss at about 250 °C. This research suggested that half and total expansion was, respectively, reached at 260 and 350 °C. In the literature, it is suggested that expansion occurs via the sulfuric acid decomposition.

The addition of EG into the polymer matrix (Polyvinyl chloride (PVC), polylactic acid (PLA)) improves thermal stability, both in an oxidizing atmosphere and in an inert atmosphere [32]. Graphene influences the thermal properties of polyethylene terephthalate (PET). An increase of graphene at 0.025% affects the thermo-oxidative resistance of the polymeric material. The addition of expanded graphite does not affect the melting point and glass transition temperature of PET [33]. Additionally, it has been observed that the addition of nanofillers does not affect the crystallization temperature, nor the degree of crystallinity of PET. In an oxidizing and inert atmosphere, the thermal degradation process of PET/EG nanocomposites has shown improved thermal stability. The mechanism of PET stabilization by graphene is associated with the transfer of free radicals on the carbon planes. Within 2% weight loss, nanocomposites containing EG, already more than 0.1% by weight, possess higher thermo-oxidative stability.

The characteristics of EG are often influenced by the modifying agent (oxidant, intercalant, and assistant intercalant) used during the intercalation process. The influence of $KMnO_4$ (oxidant agent) dosage on EG characteristics was studied by Pang et al. [34] and Zhao et al. [26]. In this research, at the range of 0.2~0.6 g/g, the influence of $KMnO_4$ dosage was detected at the mass ratio of C: H_2SO_4 (98%): $Na_4P_2O_7$ = 1.0:5.0:0.6 (g/g), the temperature conditions were 40 °C, with the reaction time of 1 h. In addition, it was observed that H_2SO_4 should be diluted to the mass concentration of 80% before the reaction. The results showed that when the mass ratio of $KMnO_4$ to C was set to 0.4:1.0, the minimum initiation expansion temperature was 160 °C. On the other hand, the insufficiency of $KMnO_4$ could lead to incomplete oxygenation of graphite and decrease expandable volume (EV) of a product. Superfluous $KMnO_4$ would cause excessive oxygenation of graphite, leading to a decrease of EG granularity and then reduced EV. In the tested dosage, the increase of $KMnO_4$ would cause an increase in initiation expansion temperature. The feasible dosage of $KMnO_4$ could be set as 0.4 g/g. Similarly, in the same work, the influence of H_2SO_4 dosage on EG characteristics was analyzed. In the range of 3.0~6.0 g/g, the influence of H_2SO_4 dosage with a mass concentration of 98% was detected at the mass ratio of C:$KMnO_4$:$Na_4P_2O_7$ = 1.0:0.4:0.6 (g/g) at 40 °C reacting for 1 h, and H_2SO_4 was diluted to the mass concentration of 80% before reaction. Results proved that when the mass ratio of H_2SO_4 to C was controlled as 5.0:1.0, EG had the lowest initiation expansion temperature of 165 °C and a maximum EV of 550 mL/g. Insufficiency H_2SO_4 would cause an incomplete intercalation reaction, leading to a decrease of expansion degree. Superfluous H_2SO_4 would cause the relative scarcity of $KMnO_4$ and incomplete oxygenation of graphite. In the tested dosage, an increase of H_2SO_4 could cause V type changes of initiation expansion temperature; the lowest initiation expansion temperature of 165 °C could be gained with EG possessing maximum expansion volume. The feasible dosage of H_2SO_4 could be set as 5.0 g/g.

In some cases, EG will work synergistically with other additives to improve the flammability properties of polymers. The synergistic FR effect between EG and modified ammonium polyphosphate (M-APP) was investigated in the case of wood floor-polypropylene composites [35]. The results from the cone calorimeter demonstrated that both EG and M-APP could effectively improve the flame retardancy of Wood Plastic Composite (WPC). The function became stronger as the EG content increased. The sample of EG:M-APP = 1:1 did have longer burning time, with lower heat rate release (HRR), total heat release (THR), smoldering test (TSP), and mass loss value. Limiting oxygen index (LOI) results of the same sample (EG:M-APP = 1:1) reached the best value. The thermal degradation analysis showed that the stability was improved with the addition of EG and M-APP, especially at the proportion of 1:1. However, the mechanical properties test results showed that EG deteriorated the mechanical properties of WPC, but when the EG:M-APP = 1:1, the mechanical properties did not decrease obviously, and flame retardancy was better than for M-APP/WPC.

As stated before, the characteristics of the char layer are significant to better FR properties of polymeric materials. The effect of EG with classical intumescent ingredients, to infer the residual weight and structure of char of EG-APP-Mel-boric, was investigated [36]. In this research, the expansion of the char increased by increasing the weight of EG at a certain limit. SEM result showed that the structure of residue char was improved by increasing the wt% of EG. X-Ray Diffraction (XRD) and Fourier-transform infrared spectroscopy (FTIR) analysis showed the presence of boron oxide and boron phosphate in the residual char. Thermogravimetric analysis (TGA) and derivative thermogravimetric analysis (DTGA) showed that EG could enhance residue weight higher than that of APP-melamine-boric acid-epoxy-hardener coating. EG is a carbon source that will create a uniform protective layer on the surface of the insulating materials. Thus, the efficiency of the heat transfer could be reduced, and this would give a better effect of intumescent into the coating formulations [36].

A new application for EG as IFR was worked out by Batista et al. [29]. The objective of this work was the cellulose-based model material (CMM) encrusted with EG. The general aim of the research was to determine its basic fire resistance properties. The scope of the research involved measurement of the following parameters: time to ignition (Ti), time to flame out (Tf), heat release rate (HRR), and mass loss (ML). Oxygen index (OI) was also part of measurements.

The addition of EG was found to increase the flame resistance of cellulosic material. Although Ti for all CMM samples was lower than for control samples, this fact actually favored the promotion of char forming. The improved physical characteristics of char were achieved by increasing the amount of the insulating layer and reducing crack formation. This aspect allowed the combustion process of CMM, with a suitable EG, to be much longer than the combustion process of pure cellulose. Besides, the maximum HRR for CMM encrusted with EG was significantly smaller compared to that for pure cellulose.

6. Conclusions

One crucial parameter affecting the thermal degradation properties of cellulose is its crystallinity degree. It is observed that cellulose thermal degradation starts in the amorphous regions and propagates to its more crystalline domains. The structure of graphite consists of layers in which there are coupled six-membered aromatic cyclic systems. Carbon atoms within a graphite layer are covalently bonded, while layers are bound by weak van der Waals forces. Graphite is characterized by highly anisotropic properties and excellent electrical and thermal conductivities. It is stated that the char is formed during the whole combustion process rather than at the end of the combustion stage [37]. The overall role of char shall cease as soon as the carbon glows. Although research on expandable graphite (EG) has achieved great progress, its practical application is restricted due to high sulfur content and serious pollution during production. The presence of expandable graphite (EG) greatly improves the char formation rate and quality in FR systems. Based on this literature review, it can be assumed that the incrustation of cellulose with expandable graphite may be of particular importance in the context of the thermal resistance of its amorphous areas [27].

Despite many studies, there are still many gaps in fire retardants application, where innovative solutions like an extra source of carbon (cellulose, lignin, starch, graphite, etc.) might be utilized.

Author Contributions: Conceptualization, B.M.; methodology, A.B. and W.G. investigation, B.M., A.B. and W.G.; writing—original draft preparation, A.B., B.M., W.G.; writing—review and editing, B.M., A.B. and W.G.; supervision, B.M. and W.G. All authors have read and agreed to the published version of the manuscript.

Funding: The publication was co-financed within the framework of the Ministry of Science and Higher Education program—"*Regional Initiative Excellence*" in 2019–2022, Project No. 005/RID/2018/19. The study served as the background for the "*CellMat4ever*" project supported by the National Center for Research and Development (decision DWM/POLNOR/66/2020).

Conflicts of Interest: The authors declare no conflict of interest the results.

References

1. Beard, A.; Angeler, D. *17 Flame Retardants: Chemistry, Applications, and Environmental Impacts*; Wiley: Weinheim, Germany, 2010; pp. 415–439. Available online: https://onlinelibrary.wiley.com/doi/abs/10.1002/9783527628148.hoc017 (accessed on 6 April 2020).
2. Jesbains, K.; Faiz, A. The study of bonding mechanism of expandable graphite based intumescent coating. *Res. J. Chem. Environ.* **2011**, *15*, 401–405.
3. Camino, G.; Costa, L. Performance and mechanisms of fire retardants in polymers—A review. *Polym. Degrad. Stab.* **1988**, *20*, 271–294. [CrossRef]
4. Zheng, C.; Li, N.; Ek, M. Cellulose-fiber-based insulation materials with improved reaction-to-fire properties. *Nord. Pulp Pap. Res. J.* **2017**, *32*, 466–472. [CrossRef]
5. Nazare, S. 14 Environmentally friendly flame-retardant textiles. In *Sustainable Textiles: Life Cycle and Environmental Impact*; Blackburn, R.S., Ed.; Woodhead Publishing Limited: Sawston, Cambridge, UK, 2009; Volume 32, pp. 339–368.
6. Wilkie, C.A.; Morgan, A.B. *Fire Retardancy of Polymeric Materials*, 2nd ed.; CRC Press: Boca Raton, FL, USA, 2009.
7. Camino, G.; Costa, L.; Di Cortemiglia, M.L. Overview of fire retardant mechanisms. *Polym. Degrad. Stab.* **1991**, *33*, 131–154. [CrossRef]

8. Khan, M.; Ganster, J.; Fink, H. Natural and man-made cellulose fiber reinforced hybrid polypropylene composites: Effect of fire retardants. *Adv. Mater. Res.* **2007**, *29*, 341–344. [CrossRef]

9. Feng, C.-M.; Zhang, Y.; Lang, D.; Liu, S.-W.; Chi, Z.; Xu, J.-R. Flame retardant mechanism of a novel intumescent flame retardant polypropylene. *Procedia Eng.* **2013**, *52*, 97–104. [CrossRef]

10. Wladyka-Przybylak, M. Combustion characteristics of wood protected by intumescent coatings and the influence of different additives on fire retardant effectiveness of the coatings. *Mol. Cryst. Liq. Cryst. Sci. Technol. Sect. A Mol. Cryst. Liq. Cryst.* **2000**, *354*, 449–456. [CrossRef]

11. Hao, J.; Chow, W.K. A brief review of intumescent fire retardant coatings. *Arch. Sci. Rev.* **2003**, *46*, 89–95. [CrossRef]

12. Zhuge, J. Fire Retardant Polymer Nanocomposites: Materials Design and Thermal Degradation Modeling. Ph.D. Thesis, University of Central Florida, Orlando, FL, USA, 2012. Available online: https://stars.library. ucf.edu/cgi/viewcontent.cgi?referer=https://www.google.com/&httpsredir=1&article=3173&context=etd (accessed on 6 April 2020).

13. De Araújo, M. Fibre science: Understanding how it works and speculating on its future. In *Natural Fibres: Advances in Science and Technology Towards Industrial Applications*; Springer Science and Business Media LLC: Berlin, Germany, 2016; Volume 12, pp. 3–17.

14. Shen, D.; Xiao, R.; Gu, S.; Luo, K. The pyrolytic behavior of cellulose in lignocellulosic biomass: A review. *RSC Adv.* **2011**, *1*, 1641–1660. [CrossRef]

15. Lowden, L.A.; Hull, T.R. Flammability behaviour of wood and a review of the methods for its reduction. *Fire Sci. Rev.* **2013**, *2*, 4. [CrossRef]

16. Salmeia, K.A.; Jovic, M.; Ragaisiene, A.; Rukuiziene, Z.; Milasius, R.; Mikucioniene, D.; Gaan, S. Flammability of cellulose-based fibers and the effect of structure of phosphorus compounds on their flame retardancy. *Polymers* **2016**, *8*, 293. [CrossRef] [PubMed]

17. Mazela, B.; Broda, M. Natural Polymer-Based Flame Retardants for Wood and Wood Products. 2015. Available online: https://www.researchgate.net/publication/282156040_Natural_polymer-based_flame_ retardants_for_wood_and_wood_products (accessed on 18 May 2020).

18. Shen, D.; Gu, S. The mechanism for thermal decomposition of cellulose and its main products. *Bioresour. Technol.* **2009**, *100*, 6496–6504. [CrossRef] [PubMed]

19. Emmons, H.W.; Atreya, A. The science of wood combustion. *Proc. Indian Acad. Sci. Sect. C* **1982**, *5*, 259. [CrossRef]

20. Duquesne, S.; Le Bras, M.; Bourbigot, S.; Camino, G.; Eling, B.; Lindsay, C.; Roels, T. Expandable graphite: A fire retardant additive for polyurethane coatings. *Fire Mater.* **2003**, *27*, 103–117. [CrossRef]

21. Zabel, H.; Solin, S.A. *Graphite Intercalation Compounds I: Structure and Dynamics*; Springer Science & Business Media: Berlin, Germany, 2013; Volume 14, pp. 5–6.

22. Thangamuthu, M.; Hsieh, K.Y.; Kumar, P.V.; Chen, G.-Y. Graphene- and Graphene Oxide-Based Nanocomposite Platforms for Electrochemical Biosensing Applications. *Int. J. Mol. Sci.* **2019**, *20*, 2975. [CrossRef]

23. Yang, Q.; Geng, Y.; Dong, H.; Zhang, J.; Yu, X.; Sun, L.; Lu, X.; Chen, Y. Effect of environmental regulations on China's graphite export. *J. Clean. Prod.* **2017**, *161*, 327–334. [CrossRef]

24. Kutz, M. *Handbook of Environmental Degradation of Materials*; William Andrew: Norwich, NY, USA, 2012; Volume 3, pp. 367–385.

25. Nguyen, H.D.T.; Nguyen, T.T.; Le Thi, A.K.; Nguyen, T.D.; Bui, T.P.Q.; Bach, L.G. The preparation and characterization of $MnFe_2O_4$-decorated expanded graphite for removal of heavy oils from water. *Materials* **2019**, *12*, 1913. [CrossRef]

26. Zhao, H.; Pang, X.; Zhai, Z. Preparation and antiflame performance of expandable graphite modified with sodium hexametaphosphate. *J. Polym.* **2015**, 1–5. [CrossRef]

27. Zheng, C.; Li, N.; Ek, M. Mechanism and kinetics of thermal degradation of insulating materials developed from cellulose fiber and fire retardants. *J. Therm. Anal. Calorim.* **2018**, *135*, 3015–3027. [CrossRef]

28. Krassowski, D.W.; Hutchings, D.A.; Qureshi, S.P. *Expandable Graphite Flake as an Additive for a New Flame Retardant Resin*; GrafTech International Holdings Inc.: Brooklyn Heights, OH, USA, 2012; pp. 1–6.

29. Anielkis, B.; Wojciech, G.; Bartłomiej, M. Expandable graphite flakes as an additive for a new fire retardant coating for wood and cellulose materials—Comparison analysis. In *Wood & Fire Safety*; Springer Science and Business Media LLC: Berlin, Germany, 2020; pp. 120–124.

30. Kim, J.; Oh, S.; Kim, H.; Kim, B.; Jeon, K.-J.; Yoon, S.-H. Expanding characteristics of graphite in microwave-assisted exfoliation. In Proceedings of the Scientiic Cooperations International Workshops on Engineering Branches, Istanbul, Turkey, 8–9 August 2014.

31. Seredych, M.; Mikhalovska, L.; Mikhalovsky, S.; Gogotsi, Y. Adsorption of bovine serum albumin on carbon-based materials. *C J. Carbon Res.* **2018**, *4*, 3. [CrossRef]

32. Paszkiewicz, S. Morphology and thermal properties of expanded graphite (EG)/poly (ethylene terephthalate) (PET) nanocomposites. *Chemik* **2012**, *66*, 26–30.

33. Paszkiewicz, S.; Nachman, M.; Szymczyk, A.; Špitalský, Z.; Mosnacek, J.; Rosłaniec, Z. Influence of expanded graphite (EG) and graphene oxide (GO) on physical properties of PET based nanocomposites. *Pol. J. Chem. Technol.* **2014**, *16*, 45–50. [CrossRef]

34. Pang, X.Y.; Song, M.K.; Tian, Y.; Duan, M.W. Preparation of high dilatability expandable graphite and its flame retardancy for LLDPE. *J. Chil. Chem. Soc.* **2012**, *57*, 1318–1322. [CrossRef]

35. Guo, C.; Zhou, L.; Lv, J. Effects of expandable graphite and modified ammonium polyphosphate on the flame-retardant and mechanical properties of wood flour-polypropylene composites. *Polym. Polym. Compos.* **2013**, *21*, 449–456. [CrossRef]

36. Sami, U.; Faiz, A. The effect on expansion and thermal degradation of 63um expandable graphite on intumescent fire-retardant coating composition. *Res. J. Chem. Environ.* **2011**, *15*, 944–951.

37. Sun, L.; Xie, Y.; Ou, R.; Guo, C.; Hao, X.; Wua, Q.; Wang, Q. The influence of double-layered distribution of fire retardants on the fire retardancy and mechanical properties of wood fiber polypropylene composites. *Constr. Build. Mater.* **2020**, *242*, 118047. [CrossRef]

MDPI
St. Alban-Anlage 66
4052 Basel
Switzerland
Tel. +41 61 683 77 34
Fax +41 61 302 89 18
www.mdpi.com

Forests Editorial Office
E-mail: forests@mdpi.com
www.mdpi.com/journal/forests